Management and Industrial Engineering

Series Editor

J. Paulo Davim, Department of Mechanical Engineering, University of Aveiro, Aveiro, Portugal

This series fosters information exchange and discussion on management and industrial engineering and related aspects, namely global management, organizational development and change, strategic management, lean production, performance management, production management, quality engineering, maintenance management, productivity improvement, materials management, human resource management, workforce behavior, innovation and change, technological and organizational flexibility, self-directed work teams, knowledge management, organizational learning, learning organizations, entrepreneurship, sustainable management, etc. The series provides discussion and the exchange of information on principles, strategies, models, techniques, methodologies and applications of management and industrial engineering in the field of the different types of organizational activities. It aims to communicate the latest developments and thinking in what concerns the latest research activity relating to new organizational challenges and changes world-wide. Contributions to this book series are welcome on all subjects related with management and industrial engineering. To submit a proposal or request further information, please contact Professor J. Paulo Davim, Book Series Editor, pdavim@ua.pt

Jayakrishna Kandasamy · Aravind Raj Sakthivel ·
J. Paulo Davim
Editors

Progress in Sustainable Manufacturing

Springer

Editors
Jayakrishna Kandasamy
Department of Manufacturing Engineering
School of Mechanical Engineering
Vellore Institute of Technology
Vellore, Tamil Nadu, India

Aravind Raj Sakthivel
Department of Manufacturing Engineering
School of Mechanical Engineering
Vellore Institute of Technology
Vellore, Tamil Nadu, India

J. Paulo Davim
Department of Mechanical Engineering
University of Aveiro
Aveiro, Portugal

ISSN 2365-0532 ISSN 2365-0540 (electronic)
Management and Industrial Engineering
ISBN 978-981-99-0200-2 ISBN 978-981-99-0201-9 (eBook)
https://doi.org/10.1007/978-981-99-0201-9

This Springer imprint is published by the registered company Springer Nature Singapore Pte Ltd.
The registered company address is: 152 Beach Road, #21-01/04 Gateway East, Singapore 189721, Singapore

We dedicate to our family and friends,
for their love & support
Thank You.

Preface

Sustainable Manufacturing (SM) ensures manufacturing products that are environmentally safe, economically viable, and socially benign. The current technological disruptions and pandemic outbursts have kindled the need for transforming the manufacturing process to be more sustainable. This book comprehends the progress in the field of sustainable manufacturing by addressing the challenges in transforming manufacturing processes to be sustainable and factors affecting the sustainability of the manufacturing process. This book provides a digital roadmap for ensuring sustainability in manufacturing, which helps manufacturing organizations in their transformation towards Industry 4.0.

This book is unique as it focuses on key challenging factors and strategies that ensure sustainability across the product lifecycle. This book details the role of digital technologies in ensuring sustainability in manufacturing and provides a digital roadmap for ensuring sustainability in the manufacturing.

The first chapter Introduction to Sustainable Manufacturing explains the evolution of sustainable manufacturing over the years and need for sustainability. Also, it emphasizes the impact of sustainability in manufacturing processes along with tools for measuring the impact.

The second chapter Sustainable Material Selection focuses on processes included in sustainable material selection. It analyzes the sequence of processes involved in conventional material selection to contemporary processes which focuses on sustainability.

The third chapter Sustainable Product Design for Electric Vehicles concentrates towards the shift in manufacturing industry towards e-vehicle manufacturing and challenges involved in it. It also deeply analyzes the various kind of batteries involved in manufacturing, and its life cycle using various scientific tools from the design stage.

The fourth chapter Sustainability Assessment of Organizations Based on the Orientations of Product Sustainability gives a sustainability assessment model for manufacturing organization by considering manufacturing material, product design, and manufacturing methodology changes. The best alternative was identified using combination of various tools such as Fuzzy VIKOR, ECQFD, and LCA.

The fifth chapter Sustainability in Manufacturing focuses on pillars of sustainability (Environmental, Economical, and Social). It discusses about the sustainable manufacturing practices related to proper utilization of various resources related to manufacturing and role of Industry 4.0 in sustainability

The sixth chapter Experimental Investigation of Machining NIMONIC 80 Alloy by WEDM Process via Multi-objective Optimisation Techniques: A Sustainable Approach explains the need for multi-objective optimization techniques in the wire EDM process for the selection of optimal parameters in manufacturing process to enhance the sustainability in the entire process

The seventh chapter Prediction and Optimization of Sustainable Production Processes for Automotive Components focuses on the sustainable production of powertrain using prediction and optimization in manufacturing. Utilizing proper materials leads to lesser material wastage and carbon footprint which increases the sustainability.

The eighth chapter A Brief Review of Sustainable Composites for Food Packaging Applications presents a detailed review about various sustainable composites used for food packaging applications focusing on superior biodegradability, renewability, bioavailability, and non-toxicity. This chapter details various biopolymers and their various application areas in food packaging

The nineth chapter Improving the Sustainability of Autogenous Pulsed Current Gas Tungsten Arc Welding discusses about optimized utilization of pulsed current gas tungsten arc welding considering factors such as peak current, base current, percent on-time, pulse frequency, and welding speed.

The tenth chapter An Integration of Smart Technology in Manufacturing details about the need for smart technology inclusive of Industry 4.0 and Big data analytics to enhance the sustainability in manufacturing firms.

The eleventh chapter Embracing New Digital Technologies to Ensure Sustainability in Manufacturing explains the various contemporary technologies and their importance in sustainability. The authors detailed technologies such as cloud computing, additive manufacturing, artificial intelligence, and autonomous robots were playing a crucial role in sustainable improvement aspects.

This book will be a useful resource to researchers, students, and sustainability practitioners in understanding the recent advancements in the field of sustainable manufacturing. Finally, we thank all the authors who contributed to this book and the readers who will be using this book.

Vellore, India — Jayakrishna Kandasamy
Vellore, India — Aravind Raj Sakthivel
Aveiro, Portugal — J. Paulo Davim

Acknowledgements

The editors would like to acknowledge the help of all the people involved in this project and, more specifically, to the authors and reviewers that took part in the review process. Without their support, this book would not have become a reality.

First, the editors would like to thank each one of the authors for their contributions. Our sincere gratitude goes to the chapter's authors who contributed their time and expertise to this book.

Second, the editors wish to acknowledge the valuable contributions of the reviewers regarding the improvement of quality, coherence, and content presentation of chapters.

Jayakrishna Kandasamy
Aravind Raj Sakthivel
J. Paulo Davim

Acknowledgements

Contents

About the Editors

Prof. Jayakrishna Kandasamy is an Professor in the School of Mechanical Engineering at the Vellore Institute of Technology University, India. His research is focused on the design and management of manufacturing systems and supply chains to enhance efficiency, productivity, and sustainability performance. He has published 66 journal publications in leading SCI/SCOPUS Indexed journals, 26 book chapters, and 91 refereed conference proceedings, 7 books in Springer series'. He has been awarded Global Engineering Education Award from Industrial Engineering and Operations Management (IEOM) Society International, U.S.A. in 2021, Institution of Engineers (India)—Young Engineer Award in 2019.

Prof. Aravind Raj Sakthivel is an Associate Professor in the School of Mechanical Engineering at Vellore Institute of Technology University, India. His research area is focused on agile manufacturing process and their applications in manufacturing industries, industrial systems engineering, additive manufacturing, and Industry 4.0 projects. He has published 20 international journals papers in leading SCI/SCOPUS indexed journals and 34 papers in reputed conference proceedings; he is the co-author for four book chapters and has edited one book. He is a reviewer of journals of national and international repute.

Prof. J. Paulo Davim is a Full Professor at the University of Aveiro, Portugal. He is Senior Chartered Engineer by the Portuguese Institution of Engineers with an MBA and Specialist titles in Engineering and Industrial Management as well as in Metrology. He is also Eur Ing by FEANI-Brussels and Fellow (FIET) of IET-London. He is the Editor in Chief of several international journals, Guest Editor of journals, books Editor, book Series Editor and Scientific Advisory for many international journals and conferences. Presently, he is an Editorial Board member of 30 international journals and acts as reviewer for more than 100 prestigious Web of Science journals. In addition, he has also published as editor (and co-editor) more than 200 books and as author (and co-author) more than 15 books, 100 book chapters and 500 articles in journals and conferences. He has listed in World's Top 2% Scientists by Stanford University study.

Introduction to Sustainable Manufacturing

Hrishikesh Dutta, J. Jayaramudu, Kishore Debnath, Deba Kumar Sarma, Pubali Chetia, and S. Periyar Selvam

Abstract Sustainability refers to the use of available resources in such a way that their scarcity does not affect the future generation. Nowadays, another word that goes hand in hand with sustainability is "Green". Though *sustainability* and *green* have different definitions, they are used synonymously in the context of the present scenario in manufacturing. In the manufacturing industry, the application of less polluting energy sources, recycling of raw materials, and using processes that are less harmful to the environment are some of the methods that can contribute toward achieving sustainability. Sustainable manufacturing is a concept that has been evolving continuously since its inception. To realize sustainability in manufacturing, the idea of sustainability must be accompanied by certain tools and methods along with implantation policies. One such method to maintain sustainability is to reduce the amount of carbon emission during manufacturing. This chapter discusses

H. Dutta (✉)
Centre for additive manufacturing, Chennai Institute of Technology, Tamilnadu 600069 Chennai, India
e-mail: dutta.hrishikesh42@gmail.com; hrishikeshdutta@citchennai.net

J. Jayaramudu · P. Chetia
Polymer and Petroleum Group, Materials Sciences and Technology Division, CSIR-North East Institute of Science and Technology, Jorhat, Assam 785006, India

Academy of Scientific and Innovative Research, Ghaziabad 201002, India

J. Jayaramudu
e-mail: jayaramphd@gmail.com

K. Debnath · D. K. Sarma
Department of Mechanical Engineering, National Institute of Technology Meghalaya, Bijni Complex, Laitumkhrah, Shillong, Meghalaya 793003, India
e-mail: debnath.iitr@gmail.com

D. K. Sarma
e-mail: dksarma@nitm.ac.in

S. P. Selvam
Department of Food Process Engineering, Postharvest Research Lab, SRM Institute of Science and Technology, Potheri, Kattankulathur 603203, Tamil Nadu, India
e-mail: periyars@srmist.edu.in

J. Kandasamy et al. (eds.), *Progress in Sustainable Manufacturing*, Management and Industrial Engineering, https://doi.org/10.1007/978-981-99-0201-9_1

different aspects of sustainability in the context of manufacturing. The strategies to attain the goal of achieving sustainability in manufacturing have been explained elaborately. Moreover, different approaches adopted to maintain sustainability in the field of manufacturing have been reviewed and extensively discussed.

Keywords Sustainable manufacturing · Outcome · Impact · LCA · Methods · Tools

1 Background and Rationale

These days, humans are consuming the natural resources available on the earth at such a rapid pace that it has become an alarming situation for whole mankind as it takes centuries for these resource materials to get generated. From years 1950–2005, the worldwide consumption of natural gas and oil increased by 14 times and 6 times, respectively. If this trend of increasing consumption of natural resources continues like this, then our future generation will have to face the unavoidable circumstances of the scarcity of resources. Moreover, the lifestyles of the developing society around the globe upgraded by the advanced technologies play a vital role in maintaining the limit of resource consumption. These all concerns lead to the necessity of sustainability which means utilization of the resources in such a way that it does not compromise the fate of our future generation. We, humans, must be concerned about maintaining the balance of nature keeping the fact in mind that nature may be both rigid and fragile at times. The more we exploit nature the more we are closing toward the threshold point where the ecological balance will be shattered and it will affect every specie on the planet.

The definition of sustainability was first proposed in the year of 1987 (Brundtland report) which mentioned that unwanted changes were occurring in our surroundings including the atmosphere, animal habitat, forest, water, and soil [1]. In the context of manufacturing industry, a sustainable approach of manufacturing a product encompasses the minimization of wastes that are produced during the entire life cycle of the product and the reduction of its adverse effects on the environment.

2 Why Sustainable Manufacturing?

Sustainable manufacturing has emerged as one of the most important areas of research during the last few decades. It is a concept that encompasses the design of manufacturing processes and products along with their planning and control which lead to the identification, quantification, assessment, and management of the manufacturing waste flow to the environment with the ultimate aim to reduce the impact on the environment while also giving efforts in maximizing the efficiency of the available resource on the planet earth.

Fig. 1 Possible outcome of sustainable manufacturing

The development of a country vastly depends on its manufacturing sector which creates numerous opportunities of employment, enhances the standard of living, and adds to the growth of the country economically. Therefore, manufacturing is regarded as the driving factor for the prosperity and welfare of a society. But, at the same time due to its dependency on natural resource consumption and uncontrolled generation of wastes and gaseous emissions, manufacturing creates concerns for the very existence of our environmental balance. The direct and indirect contribution of manufacturing to the depletion of the resources available in the nature leads to deteriorating the effect on the health of human and other living beings and most importantly it affects the ecosystem. Therefore, there is an urgent need to understand and implement the concept of sustainability in the manufacturing sector. The motive of sustainable manufacturing is to make products using the minimum amount of energy and available resources. Sustainability must be exercised not only during the production stage but throughout the whole life cycle of the product.

Today, from the pen to the airplane, everything needs to be manufactured. In fact, it is impossible to imagine a world without manufacturing industry. For every manufactured product, we need resources and energy and with each gram of resource consumption, there comes the concern over the limited natural resource that is available on earth. Therefore, it is very important to realize sustainability in the context of manufacturing throughout the whole life cycle of the product. Sustainability can enhance the efficiency of the manufacturing processes by cutting down the cost that is involved in resource and energy consumption and also by reducing the wastes involved in a manufacturing process. A manufacturing industry can also avail the benefit of less costs involved in different environmental regulatory if it adopts sustainability at the very beginning of its setup. Other benefits of adopting sustainability in a manufacturing industry include better hiring of employees and their enhanced retention rate. In sustainable manufacturing, the emphasis is also on minimizing the number of parts that consumes resources. In other words, we need to optimize the use of materials to get the required product which demands more efficient use of the available resources. To reduce the harm that a manufacturing process can bring upon the environment, sustainable and green manufacturing is the need of the hour. Figure 1 presents the possible results of adopting sustainable manufacturing.

3 Impact of Manufacturing Processes on Environment

The most concerning effects of manufacturing processes on the environment are consumption of natural resources at a rapid pace that may lead to an imbalance in nature and also disturb the demand and supply chain, rise in the emission of greenhouse gases leading to unwanted changes in the climate worldwide, and increase in the industrial waste resulting in environmental pollution. There are many tools for assessing the environmental impact of a manufacturing process or a product, carbon footprint analysis, the impact equation, and life cycle assessment being a few of them. Manufacturers around the globe have started understanding the importance of sustainable manufacturing and hence, efforts are being made to counter the issues related to environmental impact of the manufacturing industry. But the harsh truth is that even if manufacturers succeed in reducing the amount of substances that are potential to pollute the environment, it will be a long way to arrive at sustainability.

3.1 Carbon Footprint Analysis

The factor that contribute mostly to global warming and climate change is the level of emission of carbon and its compound [2]. The rise of the automobile section of the manufacturing industry has immensely contributed to the emission of greenhouse gases (GHGs) and this is true for the entire world [3]. Carbon foot prints (CF) represents the amount of carbon emission that are potentially polluting the atmosphere during every operation that is performed on a product starting from raw material to its disposal. Carbon footprint analysis (CPA) involves the process of determining the quantity of GHGs produced during the life cycle of a manufactured product and it is presented as equivalent carbon dioxide (eCO2) [4]. The protocol that is to be followed regarding emission of GHGs was developed by World Resources Institute and World Business Council for Sustainable Development and it is considered as the prevalently used standard for the measurement of emission of carbon worldwide [5]. In recent times, the concept of CPA has become popular worldwide among the general section of people in context of their responsibilities in prohibiting global warming [6].

So, we now understand that CPA is an analytical method to quantify the effect of carbon emissions including GHGs produced by various industries or organization. For any organization, the emission of GHGs can be categorized as—(a) direct emission, (b) indirect emission, and (c) emission due to electricity [7].

On a global basis, the production sector is responsible for 20% of overall carbon emissions which consumes 54% of the total energy worldwide from various sources. Product carbon footprint (PCF) is a concept where the total emission of GHGs is measured for a particular product and it is considered as the very initial step towards minimizing CF in a manufacturing industry [8]. Some of the most emitted GHGs are briefly explained here:

(a) Carbon dioxide (CO_2): CO_2 emissions generally result due to burning of solid waste, fossil fuels, and products made of wood and tree. It can be also observed in the chemical reactions involved in manufacturing of cement.
(b) Methane (CH_4): This gas is released while producing and transporting natural gas and coal. It can also be emitted from landfill decay, livestock, and agricultural activities.
(c) Nitrous oxide (N_2O): this type of emission is the result of emissions result of industrial and agricultural activities. Combustion of solid waste and fossil fuels is also a reason for such emission.

3.2 *The Impact Equation*

There is another technique to assess the impact of technology on the environment which is known as the impact equation [9]. The main components of the impact equation are (i) population (P), (ii) affluence (A) measured by gross domestic product per person, and (iii) technology (T) measured as impact per unit of gross domestic product. As much cannot be done about population growth and also people cannot be discouraged from aspiring for more affluence, the only factor responsible for reducing the impact is technology.

3.3 *Life Cycle Assessment*

In order to implement the concept of sustainability in manufacturing emphasis should be given to the life cycle assessment (LCA) of the product. LCA is a method applied to analyze and quantify the impact of a process, product, or activity on the environment. In the beginning of civilization, human used to believe that there was an inexhaustible amount of resources on the earth and they had little idea of what consequence their daily activities can bring upon the environment. Later, people started to realize the limitation of the quantity of available resources that can be used to meet their needs. Also, the pollution created by the wastes produced in industries started being noticed by people and they started understanding its menacing effect on the environment.

With time, different tools to study and assess the impact of a product or a process on the environment got developed by scientists and researchers, LCA being one among them. LCA is used to study the overall impact of any process or product on the environment which is the sum of the different stages throughout the whole life cycle. According to ISO 14040-14,044, the goal of LCA is to compare the environmental performance of products in order to be able to choose the least burdensome. The term "life-cycle" refers to the notion that for a fair, holistic assessment the raw material production, manufacture, distribution, use, and disposal (including all intervening transportation steps) need to be assessed. The concept can also be used to optimize the environmental performance of a single product (Ecodesign) or that of a company as per ISO 2006 [10].

LCA, in fact, is one kind of analysis tool used for the evaluation of the potential effect of a material or a product on the environment or the ecological system to be precise, right from the production stage to its end of life. It is a tool which is accepted internationally from the perspective of environment as it is a holistic approach having standardized and scientific methodology. LCA facilitates an assessment of goods, products, and services which is systematic and quantitative, in terms of impact on environment, human health, and consumption of resources. In Fig. 2, the four steps of LCA study according to ISO 14040 are shown. All the inputs and outputs of the boundaries of the system are needed to be quantified and they should be assigned to the various processes and stages of the products. The typical inputs are raw materials and energy and typical outputs are emissions (air, water, and soil) and wastes. The basic steps to be followed while performing LCA (according to ISO: 14,044:2006) are presented in Fig. 2.

4 Methods and Tools to Maintain and Assess Sustainability in Manufacturing

Although, by now, the importance of sustainability in manufacturing has been understood all over the globe, there must be some well-designed systematic approaches or methods that can help a manufacturing industry or organization to fathom the environmental impact of any manufactured product or the processes involved in manufacturing it and provide a solution to improve its environmental performance. These methods and tools help the organization to understand the way the product interacts with the environment. These tools are generally termed environment management tools. The determination of the effectiveness and wastage resulted from a product by considering the inputs given to manufacture it along with the plausible output is also important. Moreover, the cost consideration is also an aspect to be considered while using these tools. The total cost related to a product consists of the costs related to acquiring raw materials, labor cost, cost of using energy, and cost of waste disposal. Product life cycle of also an important aspect with respect to its environmental impact. It is important for the manufacturers, as well as the users, to understand that short

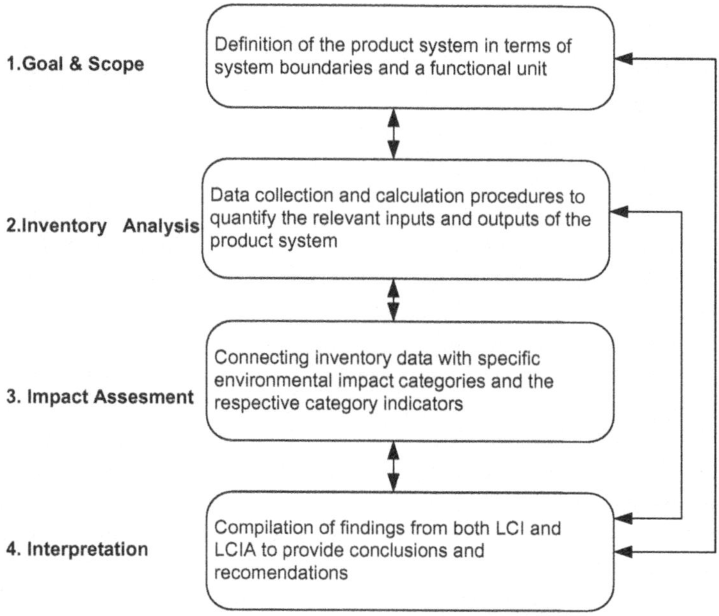

LCI: Life cycle inventory
LCIA: Life cycle inventory analysis

Fig. 2 Steps for LCA study according to ISO: 14,044:2006

life cycle of a product leads to a reduced adverse impact on the environment. The processes required to get the desired product can be divided into three phases. The first phase is the conceptualization of the required product and then designing it in detail. Then, comes the production or manufacturing phase. The last phase is the operational use and system support.

In the manufacturing sector, there is a concept of three "R"s to maintain sustainability and those are **R**educe, **R**euse, and **R**ecycle.

Reduce:

(a) We must reduce using non-degradable disposable products.
(b) We should also try eliminating the excess use of materials for packaging.
(c) Using soft copy (email) instead of hard copy (paper) when exchanging information or data is another important thing to be considered.
(d) We should opt for buying products that are durable and repairable.
(e) Increase the efficiency of the raw materials while using them.

Reuse:

(a) We should think about how we can reuse waste/scrap materials.
(b) Try to repair the old instrument or machine rather than buy a new one.

Recycle:

(a) We should accustom us to use recycled materials such as plastics, metals, and papers.
(b) Use of hazardous materials or products should be avoided.

Studies on sustainability reveal that there is a diversity of various approaches for developing sustainability assessment tools. In the study carried out by Feng and Joung, [11], an analysis of 13 mostly discussed indicators that are used for the assessment of sustainability was performed. It was concluded from the study that most of those indicators focused on only the externally developed reports and not on the internal information that can be used by the decision-makers. In another study [12], researchers carried out an analysis of eight different tools for the assessment of sustainability and found that some of those tools focused more on the assessment of the product rather than the adopted manufacturing processes. A few of the other tools helped in assessing the environmental characteristic of the product. Some other factors that are used to measure sustainability are index for sustainable development of composite [13], indices for sustainable production [14], index for sustainable manufacturing [15], etc.

Marconi and Menghi [16], in their article, proposed a tool and method for sustainable manufacturing. The application of the tool was discussed in the context of manufacturing industry that dealt with components of automobiles. The proposed too worked on the process of mapping of the activities involved in manufacturing a product with the used resource materials. The method was based on Resource Value Mapping (RVM) as explained by Papetti et al. [17]. The first step of the RVM method is defining the goal consisting of all the data and information required for the subsequent steps. In the second step which is basically mapping of the processes, it is required to analyze the production system in order to finalize the layout for plant or production line. All the important data are collected in the third step. The fourth step involves assessing the key performance indicators of the system and their representation in simple and easy-to-interpret ways. The last step is about identification of the scope of improving the ways of mitigating the complexities recognized in the previous step and also, verification of the potential efficiency, environmental and economic benefits.

5 Degree of Sustainability

The *degree of sustainability* may be defined as the extent to which a process or product is made green or sustainable. For example, if a person is using a product that is solely made from a bio-based and biodegradable polymer, it can be considered that the product has the highest degree of sustainability or degree of greenness. The reason is that using a bio-based and fully biodegradable polymeric product results in two positive outcomes. First, it totally avoids using synthetic polymer which is made from fossil fuel and minimizes the consumption of the available resources. Secondly, being fully biodegradable, it eliminates the adverse impact on the environment totally.

There are certain products that are partly sustainable. Let us take an example of a product which is made of a composite material containing biodegradable polymer as the matrix phase and a ceramic powder as the reinforcement or the filler materials. Now, after the useful life of the product, when it is disposed of for biodegradation, it cannot degrade fully due to the presence of non-biodegradable ceramic content. Therefore, it can be said that this product was partly green or sustainable or the degree of sustainability is less than that of the previous example. There is another category of products or materials which are totally non-biodegradable, for example, single-use plastic bags or plastic water bottles. These materials can never degrade naturally and therefore another approach such as incineration is taken to degrade them which may eventually create more hazards for the environment. These types of products are said to have zero degrees of sustainability.

6 Conclusion

This chapter describes the role of sustainability in manufacturing sector. Different ways to assess the impact of manufactured products or processes were explained extensively. The most important factors for assessing the impact of manufacturing on the environment are carbon footprint analysis and life cycle assessment. There are different methods of maintaining sustainability in an organization, particularly in a manufacturing industry. Also, different tools are adopted to assess the sustainability or greenness of a product or process. The concept of reduce, reuse, and recycle is the need of the hour for every manufacturing around the globe in order to achieve sustainability. With the understanding of the approach of sustainability, one must be aware of the degree of sustainability that a product exhibits. Finally, it can be said that survival of mankind and all the creatures on earth depends greatly on the fact that it's time we chose a greener and sustainable lifestyle.

References

1. Brundtland GH (1987) Our common future. Oxford University Press, New York, NY
2. Zhang T, Yunsong Y, Zhang Z (2018) Effects of non-aqueous solvents on CO_2 absorption in monoethanolamine: ab initio calculations. Mol Simul 44(10):815–825
3. Fumo N, Biswas MAR (2015) Regression analysis for prediction of residential energy consumption. Renew Sustain Energy Rev 47:332–343
4. Singh R, Chandrasekar V, Nambi VE, Arora S (2018) Carbon foot print analysis: a tool for sustainable food process development
5. Wiedmann T, Barrett J (2011) A greenhouse gas footprint analysis of UK Central Government, 1990–2008. Environ Sci Policy 14(8):1041–1051

6. Fry J, Lenzen M, Jin Y, Wakiyama T, Baynes T, Wiedmann T, Malik A, Chen G, Wang Y, Geschke A, Schandl H (2018) Assessing carbon footprints of cities under limited information. J Clean Prod 176:1254–1270
7. Ghosh P, Jha A, Sharma RRK (2020) Managing carbon footprint for a sustainable supply chain: a systematic literature review. Modern Supply Chain Res Appl
8. https://www.weforum.org/impact/carbon-footprint-manufacturing-industry/
9. Domfeld D (2014) Moving towards green and sustainable manufacturing. Int J Precis Eng Manuf-Green Technol 1(1):63–66. https://doi.org/10.1007/s40684-014-0010-7
10. Umair S (2006) Environmental impact of fiber composite materials. Royal Institute of Technology, Department of Urban Planning and Environment
11. Feng SC, Joung CB (2009) An overview of a proposed measurement infrastructure for sustainable manufacturing. In: Proceedings of the 7th global conference on sustainable manufacturing, vol 355, Chennai, India, p 360
12. Rosen MA, Kishawy HA (2012) Sustainable manufacturing and design: concepts, practices and needs. Sustainability 4(2):154–174
13. Veleva V, Ellenbecker M (2001) Indicators of sustainable production: framework and methodology. J Clean Prod 9(6):519–549
14. Krajnc D, Glavič P (2005) A model for integrated assessment of sustainable development. Resour Conserv Recycl 43(2):189–208
15. Moneim AFA, Galal NM, Shakwy ME. 2013. Sustainable manufacturing indicators. In: Global climate change. Biodiversity and sustainability, Egypt
16. Marconi M, Menghi R (2021) A sustainable manufacturing tool for the analysis and management of resource consumption within production processes. Int J Interact Des Manuf (IJIDeM) 15(1):65–68
17. Papetti A, Menghi R, Di Domizio G, Germani M, Marconi M (2019) Resources value mapping: a method to assess the resource efficiency of manufacturing systems. Appl Energy 249:326–342

Sustainable Material Selection

Yashwant Singh Bisht

Abstract Sustainable manufacturing is defined as the production of manufactured goods using cost-effective procedures that reduce negative environmental consequences while preserving energy and natural resources Sustainable business strategies are bringing significant financial and environmental benefits to an increasing number of manufacturers. Employee, community and product safety are all improved by sustainable production. With finite resources, the growth of complicated products and services has put a strain on our environment, leading to an increasingly unsustainable lifestyle. These are primarily brought on by overpopulation, Overconsumption, Resources being limited. Pollution Scientists and politicians have a poor grasp of the impact on our ecosystem and world due to a lack of or restricted habitat restoration. A growing number of businesses are included 'sustainability' as a fundamental goal in their strategy and operations to increase growth and global competitiveness. This movement has grown to encompass many well-known organizations from many industries, far beyond the small group of corporations that have traditionally positioned themselves as 'green'. In a variety of situations, these measures are having a significant impact. It covers many well known industries. These initiatives are yielding major effects in several circumstances. Automakers are attempting to decrease the number of different types of plastics used in autos to four or five to make recycling easier. This will be a difficult engineering challenge. Automakers are attempting to decrease the number of different types of plastics used in autos to four or five to make recycling easier. This will be a difficult engineering challenge. With finite resources, the growth of complicated products and services has put a strain on our environment, leading to an increasingly unsustainable lifestyle. These are primarily brought on by Overpopulation and Over-consumption Resources are limited. Pollution Scientists and politicians have a poor grasp of the impact on our ecosystem and world due to a lack of or restricted habitat restoration. Automakers are attempting to decrease the number of different types of plastics used in autos to four or five to make recycling easier. This will be a difficult engineering challenge. Take, for example, plastics, which are made from a variety of raw ingredients.

Y. S. Bisht (✉)
Department of Mechanical Engineering, Uttaranchal Institute of Technology, Uttaranchal University, Dehradun, India
e-mail: yashwantbisht@uttaranchaluniversity.ac.in

J. Kandasamy et al. (eds.), *Progress in Sustainable Manufacturing*, Management and Industrial Engineering, https://doi.org/10.1007/978-981-99-0201-9_2

There are fifty-eight different families of plastics, with over one thousand different grades, many of which are specialized for specific applications. The design and development of products with a low environmental impact are known as sustainable product development. This necessitates a comprehensive approach that examines a product's whole development and life cycle employing analysis (LCA). Design criteria for selection lowering the number of raw materials cut down on the number of parts lower the energy consumption of the products lengthen the useful life cycle use as many renewable and recyclable materials as possible evaluate and reduce the product's environmental impact across its full life cycle Technologies that make it possible Design as a criterion for selection lowering the number of raw materials FEA stands for finite element analysis. Cut down on the number of parts DFMA stands for Design for Manufacturing and Assembly. Rapid prototyping for modelling lowers the energy consumption of the products and increases the useful life cycle through thermal analysis. Life cycle testing Analysis of Failure Modes use as many renewable and recyclable materials as possible evaluate and reduce the product's environmental impact across its full life cycle Cambridge Materials Selector Software Okala Ecodesign (Ecological concerns). There are several raw materials to pick from, and the list continues to grow. Materials selection is usually a compromise between opposing factors. As a result, it's critical to remember a few key points: all materials are recyclable, though some are more difficult to recycle than others; reuse and recycling of raw materials reduce energy consumption, pollution and human injury; design for disassembly must be an integral design consideration; material selection must include a complex set of considerations and proper eco-labelling, such as recycling codes, is beneficial. There are several raw materials from which to choose, and the list continues to grow. Regardless, designers and the general public must recognise that, in addition to functional considerations, material selection must be done with environmental impact in mind.

Keywords Methods · Models · Types · Characteristics · Selection criteria · Benefits

1 Introduction

When we look back a decade or two, we can see an immense technological advancement in the industrial sector. This advancement has not only boosted the revenue of the industrial sector as each and every sector has also gained major advancement and because of this sudden giant leap in technology and advancement we have skipped the twentieth century and have entered the twenty-first century after the nineteenth. This is a very important factor for the industries as because of this they are now practising smart and innovative manufacturing and are able to get high revenue and profits. This advancement in manufacturing helps in many ways such that it helps to bring new methods and ideas for manufacturing and one such method of manufacturing is sustainable manufacturing.

When we discuss about sustainable manufacturing then in simple words, we can say that sustainable manufacturing is the production of goods through those processes which do not harm the environment and help in conserving the energy used for manufacturing and also it helps in creating less wastage of raw materials [1]. Due to this application many organizations are able to make high income and good profits, also in today's day and age many industries are practising the method of sustainable manufacturing as it requires lesser amount of raw materials, lesser energy for production which then brings down the cost of investment and also there is a lot less wastage of materials while a product is being manufactured and all of this in turn helps to increase the profits of the organization and the salaries of employees which helps in creating a good working environment [2].

Also, the companies that are practising the trend of sustainable manufacturing call themselves as 'GREEN' and this trend now is being well followed by other companies in the market. The 'GREEN' world here simply refers to the less harm that a product is causing to the environment by not wasting so much raw material and also by conserving energy to its maximum extent and not wasting it in the course of its production [3]. One must ask that why all these steps are being practised by big and small companies all over the globe even after they have already reaped good profits before, this is because now even the customers are aware and have become highly educated. Therefore, these days customers are choosing to buy products which are 'GREEN' because they think that these types of products are environment friendly and ultimately these products will cause less harm to the environment in some way or the other. Also, now people are willing to help save the environment in any way possible as now they know that harm to the environment is an ultimate harm to mankind [4]. Sustainable manufacturing has several other benefits and one of the most important ones is less wastage or usage of raw materials.

In the nineteenth century, industries did not care so much about raw material because they had plenty and they did not know that some of these raw materials can be exhausted from the earth if used lavishly. But in today's industrial age almost all the industries know and are aware of the state of the natural resources that are left on the earth and therefore important steps are being taken to reduce their wastage so that they may be there for future generation to use [5]. Also, in this industrial age a very common but powerful phrase is used and that is 'work smarter not harder', Industries these days are relying on working smarter in this age as it saves the energy of their employees and also helps in reducing the total energy usage be it electricity or in some other aspect and it greatly helps in increasing the revenue of the company [6]. All of this comes down to show us that there are many benefits of practising sustainable manufacturing such as increase in efficiency of operations which then reduces the cost of production and less wastage of materials; it helps to build new and loyal customers for the company and all of this helps to build up a name and reputation of the organization [7]. The practise of sustainable manufacturing also helps in bringing innovation for the advancement in production of its products and its manufacturing, and this helps the company to focus and achieve its long-term goals. It also greatly helps in increasing the revenue and profit of the company which in turn helps the company to pay its employees fairly well and also helps to build a

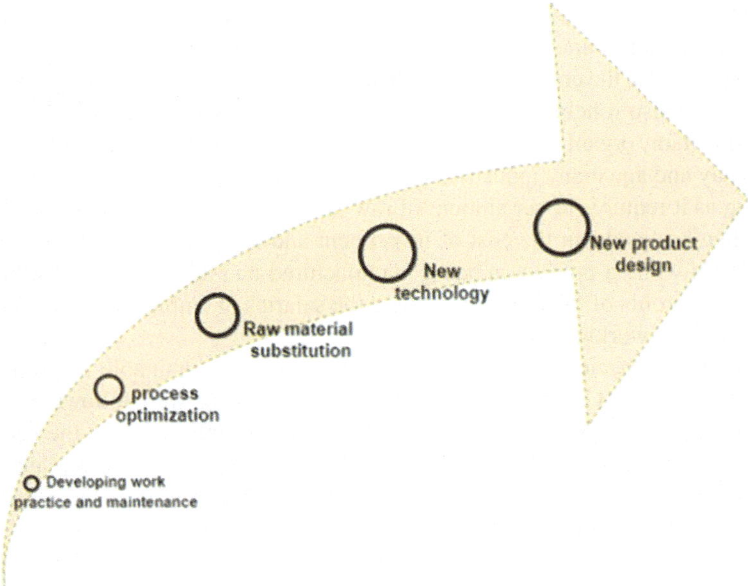

Fig. 1 Steps for new product design using sustainable material

good infrastructure for its employees while providing them with several other benefits like health insurance, loans, etc. This helps in creating a very good workspace for the employees and when a person has a good salary and a good workspace then he is very satisfied and happy and is able to work at his full potential which helps to increase the production or yield of the company to bring in more revenue and profit [8] (Fig. 1).

But even in today's day and age, there are several organizations who are not aware of the benefits of sustainable manufacturing. Also some of the organizations are aware but are simply not applying it in their industries as the process of sustainable manufacturing can be a little bit expensive due to the changes that these companies have to make in their infrastructure and as these changes also take time to be implemented therefore, these organizations are not willing to make the sacrifice due to the losses that they will face. But they do not know that the little losses that they are not ready to bear today can lead their companies to destruction tomorrow also they do not know the gains and profits that they can get by practising sustainable manufacturing today which will help in the long run of their company. Hence, sustainable manufacturing is a very important part of manufacturing and it should be practised and accepted by everyone because it provides benefits not only to 'MAN' but also to the environment where we all live and also will our generations to come [9].

2 Sustainable Material

Sustainable materials are those materials which are used by industries and they can be produced or used in the quantity required by the industries. The main point of sustainable materials is that they are used in such a way that there is no exhaustion of non-renewable resources and also, they are not used in such extent that they cause harm to the environment, in simple words they do not disturb the equilibrium of nature or the environment. Such materials in today's world are plenty such as strong fibres or polymers which can be produced in labs, also many sustainable materials have the property by which it they can be reused or recycled such as glass can be recycled again and again and does not have the need to be produced every time. The main agenda of sustainable materials is mainly to synthesize maximum number of materials in labs that can be used in production instead of the resources we acquire from the earth. By doing this there would be less and less need to extract raw materials from the earth and thus our environment's equilibrium will not be disturbed, and the non-renewable resources will not get exhausted [10]. Also, in the process of extracting raw materials from the earth there is a lot of harm to the environment and also there is a lot of pollution caused and by using sustainable materials in manufacturing this also will be reduced to a great extent (Fig. 2).

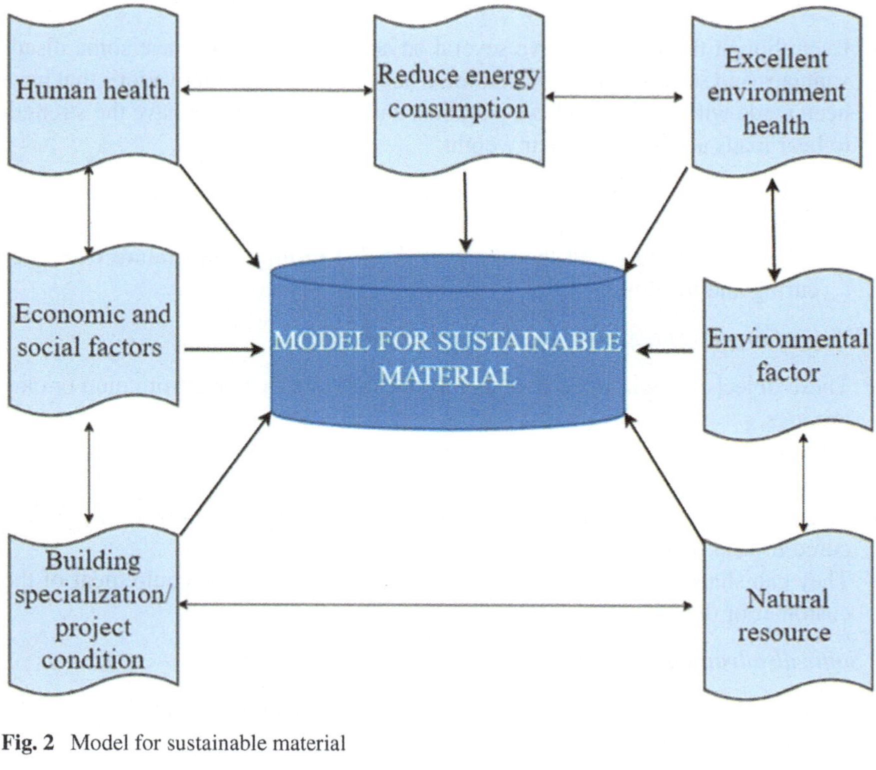

Fig. 2 Model for sustainable material

3 Types of Sustainable Materials

There are many types of sustainable materials and some of them are recycled metals, ceramics, synthetic polymers, elastomers, natural organic materials, e.g. wood, composites, etc. Now let us study some of their properties:

- **Metals**: metals such as steel, aluminium and titanium fall under the category of sustainable materials. They have several advantages over metals like cast iron which is the simplest type of metal.

 # Some of the advantages of these materials are as follows:

- Some of these metals are not so expensive therefore, they can be used widely.
- Also, they have several other beneficial characteristics like ductility, malleability, etc. and therefore they are easy to work with and ac be easily shaped and formed.
- These metals show much more resistance to corrosion as compared to simple metals like cast iron.
- Also, these metals can be made based on the characteristics required by the buyer or the clients. For example, if the buyer wants a metal with high temperature bearing property or a low melting point then such metal can be made by mixing of metals with such properties to get the desired metal.
  - These metals also have some disadvantages, and they are as follows:
- Even though these metals have several advantages, they also have some disadvantages and some of them are that there have been limited innovations that have been made with respect to these materials and also, they do not have the strength to bear loads according to their weight.
  - **Ceramics**: Ceramics are those objects which are made hard, brittle, heat-resistant and corrosion resistant in nature. They are made by shaping as in the case of pottery and then they are heated in high-temperature ovens for curing and finishing.

 # Some of the advantages of ceramics are as follows:

- These objects are non-toxic in nature as they are mostly made from mud or clay mixings.
- They are hard and are also durable in nature.
- They have resistance to corrosion which is much more than metals.
- They can withstand high temperatures while being used as they are made and cured in very high temperatures.
- They can shape into various shapes easily depending on the requirement of the customer or the client.

 # Some disadvantages of ceramics are as follows:

- There are limited number of materials from which ceramics are made.
- They are brittle in nature.

- Ceramics are also very hard to work with on machines as they are brittle in nature, and they tend to crack easily.

- **Thermoplastic polymers**: Thermoplastic is a type of plastic which when heated gets soft and when it is cooled then it gets hard. Also, these materials are easily recyclable, and their chemical properties do not change even when heated and cooled again and again. Some of the examples of thermoplastics are acrylic, polyester, nylon, etc.

 # The advantages of thermoplastics are as follows:

- They are light but also tough in nature.
- They are easy to mould, shape and can be formed and machined easily.
- They are cheap and affordable and therefore raise the standard of living.
- They also have a great resistance to corrosion.

 # Some of the disadvantages of thermoplastics are as follows:

- They have a low strength-to-weight ratio.
- Also, cannot bear very high temperatures and are hence temperature sensitive.

- **Thermoset polymer**: a thermosetting polymer is a polymer that gets permanently set once it gets hardened and the shape of these objects is set during the moulding process and therefore it cannot be softened again. Some of the examples of thermosets are Bakelite, melamine-formaldehyde polymers and polyurethane.

 # Some of the advantages of thermoset polymers are as follows:

- They are light and very tough in nature.
- They are also cheap and affordable.
- They can also be moulded, shaped and machined easily.
- These materials are also highly corrosion resistant in nature and they have a higher heat bearing capacity than thermoplastics.

 # Some of the disadvantages of thermosets are as follows:

- These plastics also have a very low strength-to-weight ratio.
- Also, these materials release toxic fumes when they are burnt and this causes great extent of air pollution.

- **Elastomers**: elastomers are rubbery materials which have a chemical structure made up of long chain-like molecules that have the property of returning to their original shape after being stretched to great extents. Some of the examples of elastomers are polyisoprene, neoprene, isoprene, etc.

 # Some of the advantages of elastomers are follows:

- They are light and tough in nature and also have very great impact bearing property.
- They can also be moulded, formed and shaped easily.

- They are affordable and cheap.
- They are also corrosion resistant in nature.
- They are undergoing constant innovations and new materials are being created.

 # Some of the disadvantages of elastomers are as follows:

- They are temperature sensitive and therefore they have a fixed temperature range between which they work.
- They also have a low strength-to-weight ratio.

– **Composites**: composite materials are those materials that are defined as the materials formed by combining two or more different material constituents macroscopically that are distinct in properties, and they do not dissolve into each other. Some of the examples of composite materials are graphite-epoxy and polyester-fibreglass.

 # Some of the advantages of composites are as follows:

- They have a very high strength-to-weight ratio.
- They also have very good resistance to corrosion.
- These materials are undergoing several innovations and have given good results and new materials are being produced rapidly.
- They have also been studied and tested, therefore can be used widely and wisely according to their requirement.

 # Some of the disadvantages of composites are as follows:

- Since these materials are expensive to produce, therefore, they are not affordable.

4 Characteristics of Sustainable Materials

- **Its effect on human body**:

We use various materials in our day-to-day lives for different purposes such as the clothes we wear, the cell phones that we use and the cars we drive. For example, let's say if a person is allergic to cotton products and does not wear them, then he/she can wear products made from organic cotton which will not harm him/her. Also, in India artificial jewellery is worn commonly, but most of the females can't wear them because it can cause infection due to some element in the jewellery hence, the jewellery can be made in such a way that it does not contain that harmful element and then it will be harmless. In such a way, there are many possibilities and applications that we can apply in the products that we use daily [11].

- **The workers**:

All these resources that we use today like diamond, gold, silver, etc. are all mined and extracted from the earth. But the working condition of the workers is very bad, most of them develop lung diseases like asthma, pneumoconiosis, etc. during the course of their work. On the other hand, if the materials can be made and produced in labs then there will be no need for mining and there will be no harm to health of the workers. Heavy equipments and machines are used in mining, and there is a probability of getting injured by these machines or equipment in comparison to a person working in a lab [12].

- **Effect on environment**:

The main advantage of using sustainable materials is that it does not cause harm to the environment as most of these materials are being synthesized in labs and they do not need vast amount of raw materials to be produced also the materials taken from the environment are taken in such quantity and way that they do not harm the environment. By doing this it shows that the equilibrium of the environment is not disturbed by them as they only borrow raw materials in a limited quantity and also there is less depletion of non-renewable resources.

- **Durability**:

Sustainable materials have much more resistance to environmental factors such as rain and sun. Let us take the example of wood. Natural wood if not treated with chemicals will not be able to withstand environmental factors such as rain for a very long time but on the other hand when we make doors and windows for our houses if they are treated properly with chemicals that are needed to withstand and resist factors like rain, sun, dust, etc. and they last much longer than natural wood.

- **Economical**:

Sustainable materials are also economical and are cheap as compared to natural materials since they are produced in a lab and are not dug out of the ground hence, they are cheaper and also are much more durable.

- **Price comparison**:

Sustainable materials are also responsible for raising the quality of life of people. It is because now the products can be synthesized and are easily available and also are cheaper when compared to the same products which were being extracted from the earth as the price of the operation is very high so eventually the products ended up being expensive, but this is not the case in sustainable materials. They are economical to produce and therefore more and more people are able to buy them and use these products to raise the standard of their lives (Fig. 3).

- **Renewable resource**:

One of the main benefits of sustainable materials is that these materials fall into the category of renewable resources in other words it means that these materials can be synthesized or can be produced again and again. This is a very beneficial as there is no need to worry as these materials will never be exhausted, and they can be used according to the needs of people. For example, Wood is a natural resource and if

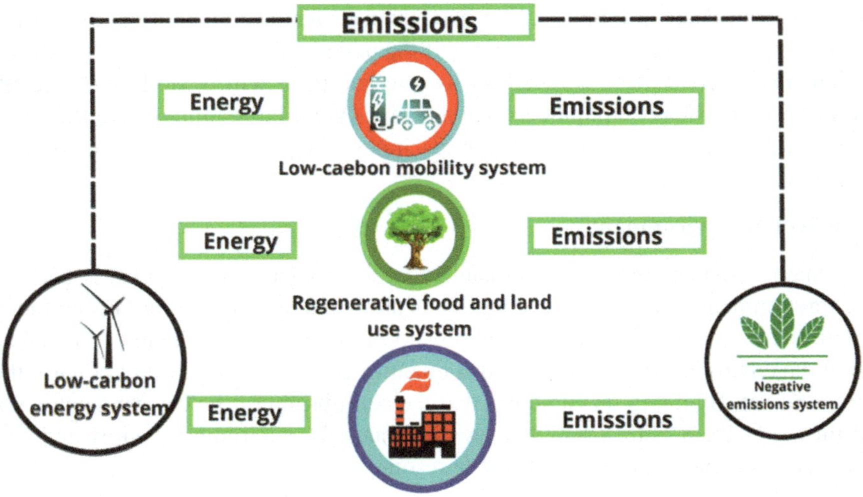

Fig. 3 Low-carbon industrial and manufacturing system for sustainable Development

used lavishly can be exhausted but on the other hand materials such as plastic cannot be totally finished as it is made in labs and can be produced several times.

- **Reuse and recycle**:

Sustainable materials have a very good property by which they can be reused and recycled. Recycling and reusing are a property by which a certain product can be recycled and can be made fit to use once again without the need of making a new product. This helps to reduce the usage of new raw materials to produce new goods every time. Also, by doing this the non-renewable resources do not deplete and there is no disturbance caused to the equilibrium of the environment. For example, glass can be recycled again and again during its life cycle and the waste of glass has worth even after it has been used but on the other hand wood ones rotten has no longer a purpose and cannot be recycled and reused is only fit for burning.

5 Method of Selecting Materials

- When we talk about the methods for sustainable material manufacturing then before that we should know what sustainability really is. In simple words, we may say that sustainability is preserving of the environment's materials for our generations to come and so that they can live peacefully and will not have to worry about some materials being finished due to misuse. Also, sustainability can be followed by the rules which are also called the 3 Es, and they are environment, equity and economics. In this the Environment is the place that we are making continuous efforts to maintain. Equity is what gives fairness when we are taking something from the environment and Economics is generally related to the well-being of the environment as well as mankind.

- To acquire sustainability or a sustainable environment, the method of selecting sustainable materials is very important as this is the process that will determine the sustainability of the material that we are choosing to work with and the material should also have the following characteristics:

1. **Cost**:

Cost is a very important factor for the benefit of the company since, if the cost of raw material will be high then the cost of the finished product will also be very high and this will not be good at all for the company as only a few selected people would be able to afford the product hence, it will not be profitable. Therefore, choosing a cheaper material is always beneficial.

2. **Ease of manufacturing**:

Ease of manufacturing can be related to the process where a material is easy and simple to work with. This is also an important factor while selecting a sustainable material as when we work on a material then we are using energy (electrical energy) through machines and various other factors and if the material will be very hard to work with then the machines will have to put more effort and more time will be required in making the material into a finished product and this is how much more energy will be used as compared to the energy used for a simpler material. Also, the technician may not be able to give his best and may get frustrated, but the main impact will be on the production line as it will be very slow as it will be hard to produce a large number of goods for the market and therefore if a low number of products reach the market, then the income and the profits will be low leading to an unstable organization.

3. **Impact on the environment**:

Whenever we are selecting a sustainable material or any material, the main impact or the main thing that gets affected is the environment. This is a very crucial decision to make while selecting a material as the environment is very important for our lives and we simply cannot survive without it. Therefore, the materials should be chosen and extracted in such or way that there should be no negative effect on the environment because if we are extracting materials from the environment without considering its equilibrium and also without considering the exhaustion of these materials then one day, we might not have any of these materials left to use or to work with which will not be good for us and for the generations to come. So, it is very important to maintain the equilibrium of the environment and not to harm the environment while extracting materials from it as our next generation and theirs will live in this environment.

4. **Fabrication and processing**:

Materials that we are choosing should have a simple form as it will be easy to fabricate and process to make them into very good products. In other words, manufacturing fabrication and processing are very important steps in the making of a product. Also, it is very important that these steps are done fast and quickly and this can only be done when the chosen material will be simple to use as this will help the worker to work easily and swiftly. This will also allow human error as in the process if the worker makes some mistake, then that will not be a big problem as the material chosen is easy to work with and another material will be able to take its place.

5. **Design**:

Design also plays an important role in the process of material selection. It is important because to make any product we have to select the raw material then only we can work on it and then we have to prepare a very good finished product for the market so that the people will buy our product and we may earn profit. But in this process the duration of making the product may take a long time depending upon the raw material that we are using. Thus, we can just make little change by selecting the material with a specific design that is according to our product and by doing this the time required for manufacturing will be reduced greatly and we will have more products to put in the market which will then help us to provide more customers with our product and with greater sale we will get greater profits.

6. **Durability**:

Durability is also a very important factor when we are in the process of selecting a material for manufacturing. It is important because when we work on a material in the industry then it must be durable and it must not be very weak and brittle as then it will be harder to work on as it will may break if the employee or the worker makes even a little mistake and then we will have to repeat the whole process again which will be a waste of time while also there will be wastage of material as we would require more material from the environment and this will cause harm to the environment. Also, these days customers have to become very aware and unforgiving and if they do not get a durable product for the money that are paying then they will no longer buy the product which will then result in bad profits for the company and also will bring down the reputation of the company.

7. **Reusability**:

In today's world reusing and recycling has become a very important factor for a company. The people in today's world are aware and they know what product to buy and most of them are attracted towards the product that can be reused or recycled. Reusability is very important and also it is a very important factor while selecting a raw material as when we select a reusable material then we are helping the environment as the same material can be reused again and again and there is not the same requirement for the material to be extracted from the environment. This also helps to reduce the usage of energy as to produce the product again and again we require more energy instead in the case of recycling and reusing, so overall it is very beneficial for the environment.

6 Selection Criteria for Sustainable Material

Selection is the process of choosing the right product for the work that is to be done, in this age selection of materials has become a little bit simple for the industries as they have already set the basic rules and parameters that the material should fall into. This helps greatly in selecting and helping companies to know that if the material is fit for their manufacturing process or not. Also, even though the process of material selection is very important to the industries but it is still a very integral part of various operations required to make a product. Thus, it is also a very important task to select the right type of material because a balance has to be created while manufacturing their products because manufacturing does not only depend upon one thing. Also, when we select these materials, it is very important for us to see if it satisfies various factors accordingly and some of them are as follows.

1. **Efficiency**:

Efficiency plays a very important role in the process of material selection as when there is no efficiency provided by the selected material in any process then there will be a lot of wastage in the process while manufacturing. Since the whole idea of sustainable manufacturing is to prevent wastage of materials and to use them to their maximum potential hence, while choosing the demanded material the responsible person should be properly specify the proportions and requirements, in the other words we should have proper idea and measurements according to what we want the materials for so that there is very less wastage of the materials. Sometimes the materials can also be designed according to the requirement of manufacturing and thus this also helps in reducing waste which ultimately brings down the cost of production as there is less material being required to manufacture and there is less material that is needed and then the equilibrium of the environment will be not disturbed and there will be lesser impact on the environment.

2. **Flexibility**:

When we are manufacturing a product, we can make mistakes sometimes during the development stages and then we have to make changes or adaptations to the existing designs while manufacturing. This is an important step to discuss as in the process of manufacturing there are many mistakes that we can make while developing a new product and also there is one great factor that we always have to watch for and that is human error. Mistakes or errors are common things when we talk about making a new product; therefore, a material which is flexible should be chosen as a raw material for production. It means that the material should have strength and other required characteristics also should have flexibility by which we can change the designs according to our requirement between the process of manufacturing and also if we want to modify our product or if we want to make any adaption design to it the material should have the flexibility and characteristics to be able to support these changes. Flexibility of material greatly helps as now we will be able to make adaptations and modifications on the basis of our requirements and there is no requirement of new material also the material should be low in maintenance.

By doing all this we will be saving a lot of raw material which will ultimately be beneficial for the company and also to the environment.

3. **Biodegradable and Recyclable**:

Suppose if we make a product and then millions of people love and purchase it then that is very good thing for the company but if that product is neither bio-degradable nor recyclable, then in this case we cannot even imagine the waste that our product will create in their environment. For example, Plastic was a revolutionary product a few years back and also it was a cheap and easy material to manufacture but it was not a bio-degradable material so that is how can we see plastics everywhere in dumps and other places even today. And even though it is a recyclable material but then the speed at which people are using and littering the environment with it is much faster than the speed at which it is being recycled and this is why there is so much plastic pollution in the world today. So, therefore, the material that we are choosing for manufacturing process must be bio-degradable and also recyclable because even after it has served its purpose it should not cause harm to the environment by causing pollution.

4. **Skill**:

In the process of manufacturing, skill is a very important factor as without skill the work performed can neither be efficient nor can it be completed according to the required standards. Hence while choosing the materials to work with, we must know and be aware that if our employees have the skillset and expertise to work with the material and then we must choose the material according to it. This is an important factor as now there are thousand types of materials to choose from and one must know and have the expertise to understand the material and should know what he is working with. Also skilled labour is also very important because in the process of manufacturing there is a lot of wastage and the skilled people are the people who help to reduce the wastage of the material and this in turn causes less harm to the environment as less material is required for manufacturing which keeps the balance in the environment while also profiting the company.

5. **Availability**:

Availability is also a very important factor for sustainable manufacturing as we all know that making of a product requires materials, and these materials are transported from different places throughout the globe. So, availability is an important factor as without the material we will not be able to produce any products and also the materials that we require for manufacturing should not be very far from the industry as the cost of transportation can add up greatly to the final cost of the product which then gets to expensive to buy and customers will then eventually not buy the product and it will create less profits for the company and can even lead it to lose.

7 Environment Benefits of Using Sustainable Materials

In today's world, the use of sustainable materials has become very important as we have to start seeing the harm that has been caused to our environment when we used its resources lavishly and now, we have to correct our mistakes and now we have to preserve it for our better future. These days almost all the industries have become aware of the sustainable materials and sustainable manufacturing therefore, more and more organizations are using these materials for making their product and using them in construction works cause using these materials not only prevent us from the harm that we are doing to the environment but also, they give us several other benefits and advantages that are helping to profit us and also the environment. Some of them are as follows.

1. **Financial improvement**:

Financial conditions play a very important role in the life of a person because it is through finances a person's living standards are determined. We can improve our financial state by choosing sustainable materials over cheap materials as sustainable materials have a very long life and they benefit the environment greatly. If a certain product on sale is not made from a sustainable material, then we must avoid buying it and instead we should buy the product made from sustainable materials as saving a few bucks and buying the cheap product is never beneficial in the long run as the cheap product will not have a longer life then the product made from sustainable materials. Therefore, we must spend a little but extra and must buy a sustainable product that works for a long time: For example, plastic water bottle should not be brought daily from a store instead we must buy a stainless steel bottle and fill water every day and use it which will have a very long life and will profit us and the environment in the long run.

2. **Better health**:

Health is also a very important factor in our lives because if a person is not healthy then what is he living for. So, we can improve our health through many factors such as using sustainable materials in our day to day to lives. Suppose let's take simple example of our house. These days we want the best looks of our houses inside and outside but that comes at a cost. Researches are being conducted these days and have also given results that the air inside the houses is much more toxic than air that is outside and this study is from the cities where there is less air pollution. This is purely because of the paints and chemicals we use to make our houses look beautiful but as they say beauty comes at a cost hence, there are several ill effects that are being caused to our health. So, instead we should use natural materials like wood to finish the inside of our houses.

3. **Reduce waste**:

We all know that when we work with a certain material, and we go on to make a finished product. In between the process there is lot of waste created and this leads to low profit and also harms the environment. Hence, we should always use sustainable materials in which the waste produced is low and also this results in profit to the company and also to the environment as less material is being used. Also, we must use materials that are reusable and recyclable as it will reduce the need to extract more raw material from the environment and will also decrease the amount of waste created as in this case the amount of waste created is being recycled and reused and it overall increases the financial state.

4. **Good feeling**:

As human we make many mistakes during the course of our lives and many of those mistakes live in our conscience throughout our lives, but on the other hand if we do some good deeds we are also comforted in our hearts, minds and we become peaceful. Researchers and Scientists have found out through conducting several experiments and studies that buying sustainable products or 'GREEN' products has a very good and positive effect on a person as by doing this we are helping the environment to sustain and grow and we are not responsible for its depletion. This effect is known as 'warm glow' and it increases the positivity of a person and it also helps and benefits us when we do well for our environment. Also, a positive and calm person is able to make many good decisions in his life which ultimately leads to his/her benefit.

5. **Longer use of materials**:

The word sustainability has the meaning that a particular material can be used for a long time. By using sustainable materials in our houses and for our construction buildings we are giving our houses and buildings a long life and also, we are creating a good environment for people to work and live in as the materials are 'GREEN' OR eco-friendly in nature. By using these materials, there is also a significant improvement to our health and also people develop more positivity as they feel good and thus this increases the productivity of a person while also it increases the profits side by side. By using sustainable materials for building, we are decreasing the overall cost in the long run as sustainable materials last very long and also, they require low maintenance and thus, the materials or the structure does not have to be changed again and again which is beneficial for the environment and ultimately beneficial to us.

6. **Less energy usage**:

When we use sustainable materials for construction, we not only are benefiting by getting a long life and less maintenance that it has but also in maximum cases when we use sustainable materials to build houses there is more insulation provided to the houses. Insulation is a very important factor as when building a house if there is poor insulation then we will require much more energy to maintain the temperature of the house. Suppose if the insulation is poor than in the summers, we will need air conditioners to cool down the house in which will in turn release chlorofluorocarbons' (CFC) and heat to the environment and CFCs are very harmful for the environment as it is responsible for ozone layer depletion which stop the ultraviolet rays from entering the earth's atmosphere, also in winter we would need much more heating than required as poor insulated homes get very cold. All of these result in the wastage and misuse of energy. Also, we should use solar panels to create energy ourselves as we have a better source of energy that is not being created by the environment and it is not going to finish anytime soon and that is the sun which also falls under the category of green energy and is very good source of energy only if we know how to use and harness it and also it reduces the overall load on the energy being created by the environmental materials. It is sustainable and even though it is expensive to setup but it has low cost of maintenance and is economical and beneficial in the long run.

8 Future of Sustainability

As by now we all know that sustainability is the process or the path that we follow which means that in sustainability we are preserving the resources for the needs of our generations to come. Sustainability has become a very important topic as to this day we have been using our resources lavishly and have exhausted some of them and also have we have depleted them greatly while also causing a great harm to the environment. In today's day and age people have become educated and aware of the situation, the availability and the importance of these resources.

People have learned that if they do not stop the abuse of the resources and of the environment then there is nothing going to be left for the future generations and who knows mankind could even go extinct because if there is no environment left, for example, the air that we breathe, the food, clothes, etc. We get it from the environment; therefore, it has become very important to preserve the environment and its resources while also use it's them cautiously. This can only happen through sustainable manufacturing that is done which sustainable materials which helps to keep the balance of resources and materials in the environment. Sustainable manufacturing uses sustainable materials that are not harmful to the environment. Also, most of these materials are recyclable and reusable; hence, there is not much need to create the same new products and this helps greatly in resources preservation of the environment and maintaining its balance. When these types of products are usable and recyclable then there is very less waste created and this helps a great deal in keeping the environment clean. So, through all this we know that sustainable manufacturing as well as sustainable materials are very important for the future as well as the present.

Also, nowadays people have realized the importance of 'GREEN' products/ sustainable products and even they are contributing greatly as they are buying and using them. They now understand why sustainability is very important for the future and one important thing to consider is that even though sometimes sustainable products might be a little expensive but on the long run they have always proven to be economical and more beneficial for us and for the environment. For example, water is a very essential part of life and therefore our bodies require daily water intake, but something that we never realize when we are outside is that, when we need water to drink, we simply buy the package water bottle and never notice the waste that we are creating by continuous usage of these plastic bottles. Even though they are recyclable but still resources and energy are needed to recycle them. We should instead use stainless steel bottles and even though they may be expensive but still they do not create waste and can be reused again and again and this is profitable in the long run.

In the same way, sustainable manufacturing is also very important for the future as in this type of manufacturing sustainable products are created and to make them sustainable materials are used. Also, in this case of sustainable manufacturing, it is done in such a way that there is less consumption of energy and also there is less wastage; hence, this is very good manufacturing process for the future as this helps in maintaining the balance in the environment while also using less resources from it. These days even construction companies are also applying the knowledge of sustainability and they are using sustainable materials for their construction works, in return the reward they get is the low cost of sustainable materials than the conventional ones and it helps to bring down the budget of the projects that they are doing. Sustainable materials greatly decrease the maintenance costs of the infrastructure, as well as providing it with a long life and also discussed before as people are buying 'GREEN' products/ sustainable products they are also helping to create a sustainable future and are also helping themselves as now they feel better about themselves as they are helping the environment in one way or other. This helps in bringing positivity in their lives and as a result positivity eventually brings happiness. All in all, sustainability is very important for the future of our environment as through sustainability only we can preserve the environment for our future.

9 Conclusion

From everything that we have discussed and said we can say that sustainability is very important for our future and sustainable manufacturing is a very important part that helps to increase and practise sustainability in this world. Also, it is only through sustainability that it will be possible to save the environment for our future generations and to undo the harms that were and are being done to the environment due to man's greedy and selfish nature and now it has become very important to start practising sustainability and also to protect the environment as it is the environment that helps us to live and be with our loved ones.

References

1. Stock T, Seliger G (2016) Opportunities of sustainable manufacturing in industry 4.0. Procedia CIRP 40:536–541
2. Bertoni M (2017) Introducing sustainability in value models to support design decision making: a systematic review. Sustainability 9(6):994

3. Hegab H, Kishawy HA (2018) Towards sustainable machining of Inconel 718 using nano-fluid minimum quantity lubrication. J Manuf Mater Process 2(3):50
4. Kishawy HA, Hegab H, Saad E (2018) Design for sustainable manufacturing: approach, implementation, and assessment. Sustainability 10(10):3604
5. Sathiskumar C, Karthikeyan S (2019) Recycling of waste tires and its energy storage application of by-products—A review. Sustain Mater Technol 22:e00125
6. Makul N (2020) Modern sustainable cement and concrete composites: review of current status, challenges and guidelines. Sustain Mater Technol 25:e00155
7. Magdy M, Alamoudy FO, Ismaeel WS (2020) An approach for sustainable material selection using life cycle assessment. Fayoum Univ J Eng 3(2):53–67
8. Machado CG, Winroth MP, Ribeiro da Silva EHD (2020) Sustainable manufacturing in industry 4.0: an emerging research agenda. Int J Prod Res 58(5):1462–1484
9. Hatefi SM, Asadi H, Shams G, Tamošaitienė J, Turskis Z (2021) Model for the sustainable material selection by applying integrated dempster-Shafer evidence theory and additive ratio assessment (ARAS) method. Sustainability 13(18):10438
10. Maheshwari PH, Waseem S (2022) Developing the processing stages of carbon fiber composite paper as efficient materials for energy conversion, storage, and conservation. In: Sustainable materials and green processing for energy conversion. Elsevier, pp 399–440
11. Mohamed HH (2022) Green processes and sustainable materials for renewable energy production via water splitting. In Sustainable materials and green processing for energy conversion. Elsevier, pp 169–212
12. Mohamed HH (2022) Introduction to green processing for sustainable materials. In Sustainable materials and green processing for energy conversion. Elsevier, pp 1–42

Sustainable Product Design for Electric Vehicles

Diwakar Suman and Sonu Rajak

Abstract Product design and development for technology plays an important role in the economic growth of the country. Sustainable products are the products that are compatible with the environment and nature throughout their entire life cycle. Traditionally, during product design, the manufacturing industry mainly focuses on the overall reduction of cost, but due to government pressure and by considering environmental problems and awareness of sustainability, manufacturers are now also considering the environmental and sustainability criteria in their decision-making process of product development. Automotive manufacturing industries need to adopt sustainability principles to survive in a globally competitive environment. Sustainable product design tools are needed to enhance the environmental, social, and economic performance of products. Researchers and practitioners have done a lot of research on product design and development, by considering the sustainability and environmental aspects need to be explored to achieve the triple bottom line of sustainability. In this context, the aim of this chapter is to categorize into three main areas of sustainability: environmental, social, and economic, and analyze with different tools for the product design of the electric vehicle. This chapter will provide a roadmap for the sustainable product design of electrical vehicles and discuss the issues, challenges, and future research direction.

Keywords Sustainable product design · Electric vehicle · Sustainability · Eco-design

---

D. Suman · S. Rajak (✉)
Department of Mechanical Engineering, National Institute of Technology Patna, Patna 800005, India
e-mail: sonu.production@gmail.com

D. Suman
e-mail: diwakars.phd20.me@nitp.ac.in

J. Kandasamy et al. (eds.), *Progress in Sustainable Manufacturing*, Management and Industrial Engineering, https://doi.org/10.1007/978-981-99-0201-9_3

1 Introduction

In the last two decades, concept of sustainability, sustainable development, and sustainable product design achieving more awareness among manufacturer and researchers [1]. Today carbon waste reduction is important for all organizations. Nowadays, researchers try to develop product design processes that are less toxic, better efficient, and safe. In recent decades, environmental issues like air pollution and global warming have been critical. With an increase in the environmental awareness, organizations have difficulties reducing the ecological impact involved in the manufacturing industry. Recently, electric vehicle (EV) concept has come to the market because they improve engineering and environmental impact [2]. In the initial stage of product design and development, designers need to use environmental aspects simultaneously with conventional design. For sustainable product design, researchers have to improve social, economic, and environmental factors like cost, capacity, manufacturing time, product yield, and pollutants. The design of electric vehicles has been given attention because of the sustainable system they want to reduce energy consumption and pollution. Electric vehicles are widely used as public transport in recent decades. The battery creates some challenges in electric vehicles because of regular and expensive maintenance and a shortage of charging facilities. Various component of the electric vehicle battery is the key component that requires proper inspection and repair for better battery life and stability. Plug-in electric vehicle battery pack standardization is used to improve sustainability. Tsang et al. [7] applied fuzzy-based product life prediction for sustainable product design in EVs. Hybrid EV gives a better alternative to fuel vehicle because they reduce air pollution and improve battery life. Kang [8] developed the relation between customer requirement and hybrid electric vehicle product design by using Kansei engineering. Kansei engineering goal is to develop and improve products by translating the customer requirement. The researcher has applied morphological analysis and a rough set to fulfill customer needs. Ultimately, vector regression is applied for optimal design with the highest Kansei value [8]. Sustainable design tools are used for products to be designed with no negative effect on environment. Although designers should consider all aspects of the product's life cycle when designing a product. Researchers discussed some tools for a separate phase in the product design process. Ahmad et al. [1] deal with the tool behavior and the paper is based on newly developed tools. The selection of tools is based on the triple bottom line of sustainability. Therefore, tool is divided into partial sustainable design tools and sustainable design tools because partial sustainable design tools do not give a triple bottom line of sustainability.

Sustainable design is necessary for the innovation of products along with decreasing the negative impact on the environment simultaneously. Sustainable design is applied to create new products and services as well as they have considered ecological, social, and economic effects from the initial product design to the final design. Three aspects of sustainable product design, environmental, social, and economic, have a link with each other. The connection of three aspects in sustainable

Table 1 Three elements to achieve sustainability in product design

Element	Explanation
Ecological aspect	Ecological design targets the reduction of negative impact on the environment from initial to final product development
Social aspect	Social aspect targets to tackle human safety. Its aim is to save industrial workers, machinery, and environment
Economic aspect	The economic aspect of product design targets the design of excellent quality innovative products with better price and cost

design reduces the negative effect on the environment [9]. The three elements to achieve sustainability in product design are shown in Table 1.

The objective of the chapter is to discuss the importance of electric vehicle, and sustainable product design with sustainability tools. In this chapter, we have to discuss different sustainability tools and compared them with each other to choose the best design tools to achieve sustainability. Several papers have been published with individual sustainable tools for sustainability and some of the researchers have integrated tools for sustainable design. Many authors have defined the sustainable design of automobile vehicle parts rather than the sustainable design of EV parts. The main area of this study is focused on the sustainable design of EVs, sustainable design of EV parts and tools. The main area of this study is focused on the sustainable design of EVs, Sustainable design of EV parts and tools. Basically, the chapter is aimed at the idea of tools for sustainable product design. Our major questions from this research area are:

- How will you classify sustainable product design tools and partial-sustainable design tools for EV?
- How can we apply sustainable product design in the field of EV parts?
- How to analyze the EV components for integrating sustainability tools?

Further, the chapter is structured as follows: An overview of electric vehicles is presented in Sect. 2. Section 3 performs the literature survey on EVs and sustainable design. Recent development based on eco-design, partial-sustainable product design (SPD), and SPD is discussed in Sect. 4. Section 5 includes future research areas and conclusion is discussed in Sect. 6.

2 Electric Vehicles (EVs)

Electric automobiles are powered wholly or partly by a source of energy in the form of battery packs and vary from typical automobiles in that they are propelled by an electric motor rather than an internal combustion engine (ICE) [19]. Electric vehicles are less expensive to operate because they have fewer moving parts to maintain, and they are also more ecologically friendly since they use little or no fossil fuels such as diesel or petrol [20]. NiMH (nickel-metal hydride) or lead-acid battery packs are used in some EVs, but lithium ion-based batteries are used and considered a standard alternative for modern-day batteries-based electric vehicles because these batteries have a longer lifespan or longevity and are excellent at maintaining energy [21]. The electric vehicles are designed in a way to customize the problems were facing by the ICEVs. Based on how far they've advanced through the manufacturing process from performance to production, the EV fleet may be categorized into four groups [22]. The many forms of electric vehicles are as follows:

A. BEVs (Battery Electric Vehicles)
B. HEVs (Hybrid Electric Vehicles)
C. PHEVs (Plug-In Hybrid Electric Vehicles)
D. FCEVs (Fuel Cell Electric Vehicles)

A. BEVs (Battery Electric Vehicles):

BEVs are also known as AEVs (All-Electric Vehicles) since they employ technology that is entirely reliant on a battery power source to power the electric powertrain, with no supplementary sources of propulsion such as fossil fuel, gasoline, hydrogen fuel cell, etc. [23]. The EVs have been produced utilizing a variety of batteries, such as NiMH (Nickel-Metal Hydrid) or lead-acid battery packs, and currently, manufacturers have expanded the range and charge storage by using lithium-ion batteries, indicating that the focus has turned to sustainability. Solid-state batteries, the next generation of batteries, are now in the experimental stage [24]. The energy is stored as chemical energy in a rechargeable battery pack, and the electricity utilized to power the vehicle is stored in a huge battery pack. The battery pack is charged simply by plugging it into an electrical outlet. Electric motor, inverter, battery, control module, and drive train are the main components of a BEV. Tesla Model 3, TATA Nexon, Mahindra E20+, Hyundai Kona, and TATA Tigor are examples of BEVs [25].

The electric motor is powered by the EV's DC battery power, which is converted to AC. A signal is transmitted to the controller when the EV's accelerator is pushed. The controller then controls the accelerated speed by changing the AC power frequency from the inverter to the electric motor, which causes the motor to create torques, causing the wheels to spin via the cog [26]. When the brakes are used or the automobile is decelerated, the engine acts as an alternator and creates electricity, which is utilized to charge the battery while driving.

B. HEVs (Hybrid Electric Vehicles):

The HEV has two engines: an electric motor-based engine and an internal combustion engine. In HEVs, there is no need to plug in to charge the battery pack; instead, regenerative braking and the ICE charge the battery pack [27]. The IC engine, gasoline tank, electric motor with battery pack, inverter and controller, and control module are the main components of an HEV. Hybrid buses are used as public transportation in many countries, while HEV vehicles such as the Toyota Prius, Kia Optima, Honda CR-Z, and others are examples of HEV cars [28]. This car uses both BEV and ICEV technology, so it can operate in either mode.

C. PHEV (Plug-In Hybrid Electric Vehicles):

Plug-in hybrid vehicles feature both an internal combustion engine (ICE) and an electric motor. It is up to the individual to choose the running engine power source, such as gasoline or diesel, or an alternative fuel such as biodiesel [29]. It also contains a rechargeable battery pack that can be charged externally as a power source. The electric motor, battery pack, inverter, control module, and battery charger are the main components of a PHEV (onboard charging options). Fuel tank, engine Porsche Cayenne S, BMW 330e, Chevy Volt, Ford C-Max, Mercedes C350e, Mini Cooper SE, Audi A3 E-Tron, and more are examples of PHEVs [30].

PHEVs may be driven in one of two modes:

- All-electric mode, which uses the battery pack for power.
- Hybrid mode, which uses both electricity and gasoline or diesel.

The PHEVs get power from the battery pack or start-up in an electric mode and may be run until the battery is depleted; once the battery is depleted, the ICE takes over and the vehicle resumes normal operation [31]. This vehicle may be charged via the engine, regenerative braking, or by connecting to an external power source.

D. FCEV (Fuel Cell Electric Vehicle):

FCEVs are powered by hydrogen, one of the most abundant substances on the planet. Although fuel-cell cars are powered by electricity, they are not the same as battery-powered or plug-in hybrid vehicles [32]. The FCEVs are filled with pure hydrogen gas from a storage tank in the vehicle. It can offer a driving range of up to 300–400 km, once it is filled. The main components are an electric motor, fuel-cell stack, storage tank for hydrogen, battery, converter, controller, etc. [33]. Miraj from Toyota, Hyundai Tucson, and Nexo from Hyundai are such examples of FCEVs [34].

The FCEV's operating premise is that a low-temperature fuel cell is employed to create electricity from hydrogen, electric energy is then either utilized to power the vehicle or saved in a storage device such as a battery pack or ultra-capacitors and in essence, energy is converted from chemical energy to electrical [35]. Water is produced as a by-product of such chemical processes.

The electrification of the transportation industry fulfills the criterion to be a more environmentally friendly alternative to internal combustion engine vehicles (ICEVs). Since the beginning of the past few decades, several efforts have been made to break into the market with electric vehicle technology. Finally, an unexpected shift can be observed in the number of sales of electric vehicles all over the globe. This chapter outlines the many components of sustainable development while discussing the few requirements that electric vehicles must meet in order to successfully compete with internal combustion engine vehicles (ICEVs).

3 Literature Review

3.1 Electric Vehicle

In recent decades sustainable design in EVs is very important in transportation because electric vehicle has better power consumption performance and it reduces air pollution or negative effect on environment. Since electric vehicles are replacing the internal combustion engine [7]. As electric vehicle can decrease the use of petroleum, it provides zero emission of pollutant gases in the environment. So many companies are developing the importance of electric vehicles in the transportation system [10]. Due to the decrease in natural resources and the increment of waste in the environment, sustainable design has come to the market worldwide [11]. Combustion engine utilizes fuel and discharges nitrogen oxides, carbon mono oxide, and carbon dioxide which are dangerous for environment. The EVs have given more attention because of less carbon emission, high power efficiency, fuel-free, and environmental consciousness [2]. Most of the researchers have applied single tool for sustainability and some people have to integrate tools to achieve sustainable design.

3.2 Sustainable Design

Environmentally conscious quality function deployment (ECQFD) is mainly used as a tool for product design in the initial stage of product development. To attain sustainability and for sustainable product design, ECQFD has been implemented by Vinodh and Rathod [3]. At present, economic expansion design and implementation play an important role. Researchers and scientists gave sustainable design concepts by developing theories and tools to enhance the environmental, engineering, and economic impact of the product life cycle. Different practitioners used different tools to convert customer requirements into engineering characteristics. Vinodh et al. [3, 4] discussed four phases of ECQFD and they integrated ECQFD, theory of inventive problem solving (TRIZ), and analytic hierarchy process (AHP) for sustainable product design of automobile components. The ECQFD is applied to involve renovation and environmental characteristics in the product design process and it comprises four phases. Phases 1 and 2 are related to identifying the main parts of EV that are essential for the environment and innovation. Phases 3 and 4 are used to examine what design modification is most effective among the alternative design option for electric vehicles. The TRIZ consists of 39 inventive principles [37]. The purpose of TRIZ was applied for innovative process design alternatives and AHP was applied for the selection of best design alternatives [4]. Integration of ECQFD and LCA has been applied by Rathod et al. [5] to achieve the sustainable product design of an EV. In this paper, ECQFD has been applied to control environmental aspects and conventional requirements, and LCA is applied for estimating the environmental effects on product design and process. In this paper, researchers have used old product parts and developed a model by integrating production, technical rank, and environmental impact. In this paper, the result obtained from the case study gives equal importance to integrating ECQFD and LCA [5]. Sustainable design and development have allowed attention recently due to increasing customer demand to achieve sustainability. In modern sustainable design, the circular economic model has focused on the aim to achieve environmental impact. The manufacturers and governments have applied circular economics in practice to achieve sustainability. Six key points of a circular economy are recycling (end-of-life treatment), design, production and remanufacturing, transportation, consumption, and collection [6]. Ahmad et al. [1] deal with the tool behavior and the paper is based on the newly developed tools. The selection of tools is based on the triple bottom line of sustainability. They have divided tools into partial sustainable design tools and sustainable design tools because partial sustainable design tools do not give a triple bottom line of sustainability.

The term design with innovative activity to choose from different ideas and possibilities is known as sustainable product design. Conventional design process is divided into four stages. The first stage deals with problem identification and planning and the second stage is conceptual design. Conceptual design stage determines the product identification and alternative idea. The third stage deals with preliminary design, they select the best alternative concept. And last one is the final stage that provides a detailed design. This stage deals with product manufacturing and maintenance. To achieve sustainability, the researcher has to include sustainability aspect in all stages of conventional design. Conventional product design concentrates on product feature, customer requirement, product quality, and product cost. However, sustainable design focuses on the whole life cycle of product for social, economic, and environmental effects. Sustainable product design provides sustainability for the entire life cycle from selection of materials, manufacturing to end-of-life and recycle. The aim of sustainable design is to minimize resource material and also reduce the discharge of flue gases to the environment along with its aim to improve the social and economic effects on the entire life cycle.

4 Recent Development

The purpose of this chapter is to review the many recently developed product design tool with respect to the triple bottom line of sustainability. Also, review the design tool and EV-based paper in the earlier published article to give better sustainability performance. The collection of the research papers is from "Web of Science", "Science Direct", "Google scholar", "Taylor and Francis" by using different keywords. Then the collection of articles was selected on the basis of their relevancy. The search process was performed with many combinations of keywords like sustainable product design, product design, product development, electric vehicle, sustainable design tool, eco-design tools, sustainable manufacturing, design for environment, and life cycle assessment tools. Based on the keywords, relevant articles were assembled from the above-mentioned source. Based on these keywords, the topics were divided into two groups: sustainability and tools of product design. Now, the final stage for selecting a research paper is based on the most relevant papers selected on the basis of two parameters such as relevancy and journal rank. Top journals that are published in this area are such as "Resources, conservation & recycling", "Computers & industrial engineering", "Journal of cleaner production", "Applied soft computing", "Sustainable production and consumption", and "International journal of sustainable engineering".

4.1 Detailed Analysis of Literature

Various studies have been focused on by researchers in the field of sustainable product design of EV and automobile vehicle parts. Many researchers have tried to focus on battery recycling and EV charging infrastructure in different countries. Some scholars have developed the literature reviews on the main area of EVs such as charging infrastructure, battery recycling, and battery swapping. Some of the researchers have discussed recycling EV battery materials.

4.1.1 Sustainable Design Tools and Issues

The environment was the only factor in the early stage of sustainable product design. So, environmental and eco-design was implemented and many numbers of eco-design tool had developed to be applied in the early stage of product design. Previously focused much research has focused on the eco-design tool. Due to the introduction of social and economic aspects in product design, sustainable product design concepts give more attention in the current time. To fulfill this gap, this paper aims to differentiate between eco-design tools and sustainable design tools. Actually, researchers have been interested in studying the important areas in the field of sustainable design and design tools but they did not give a better understanding of sustainable design tools, partial SPD, SPD, and eco-design tools.

4.1.2 Eco-Design Tool

There are many environmental tools have been developed. Eco-design and design for environment is a basic-level design tool. Eco-design tool is basically used in Europe and design for environment is used in the United States [11]. Actually, some of the eco-design tools are very comprehensive and give a detailed solution to improve sustainability. Apart from some tools, an eco-design tool performs a preliminary quantitative analysis to improve the design criteria. Devanathan et al. [36] have discussed and classified 30 eco-design tools into three parts, quality function deployment-based tool, life cycle assessment, and checklist-based tools. This article uses qualitative and quantitative criteria to classify. In the quantitative method, a large amount of product data is required for product development process and it enters the design process at the late stage. Qualitative method is relatively simple in the early stage of product development process. Life cycle assessment is mainly a category of quantitative tools and quality function deployment is the category of qualitative tools.

4.1.3 Partial Sustainable Design Tool

The impact on the environment was considered after conventional product design approaches has been applied. In recent years, the demand for sustainable product design tool is different from the previously focused tool. Then the transformation of tool from eco-design tool to a more comprehensive tool that includes environmental factor with social or economic factor. These tools are called partial sustainable design tool because they did not fulfill all dimensions of sustainability. Previously, most of the researchers have already discussed some of the partial-SPD tool has been only depending on cost and quality parameters without the aspect of sustainability. Quality parameter has been part of traditional product design which gives only economic gains in product design. Thus, some of the expended version eco-design tools were considered as a partial-SPD tool. Types of tools used tool categories, and the different types of problem considered in the EVs problem is shown in Table 2. Majority of the partial-SPD tools were dependent on QFD. Most of the partial-SPD tools were integration of different tools from other fields and it was also common in SPD tools. In this table, most of the partial-SPD and SPD tools are quantitative in nature and only quality function deployment is qualitative in nature. Here, we have seen that mostly partial-SPD tools were used for the early stage design like problem definition, conceptual design but SPD tool like life cycle sustainability assessment were used on detailed design and later design stage.

4.1.4 Sustainable Design Tool

Sustainability is the process in which the social, environmental, and economic concerns should be addressed simultaneously in the product design process. The term partial sustainable design tools is considered limited economic analysis. Similar to partial-sustainable design tool, SPD tool was also the integration of different tools from several fields. SPD tools give a triple bottom line of sustainability because they fulfill all three aspects of sustainability to the environment. These tools are future tool because most of them gives the idea of early-stage design and development. Partial-SPD tool is based on the QFD tool but these tools give the triple bottom line of sustainability and these tools are mainly used in industry. The types of sustainable design tools used for the different types of problem considered in the EVs problems are shown in Table 3.

Table 2 Categorization of eco-design and partial-sustainable product design tool for electric vehicle

Tool	Tool categories	Description	Analysis	Design stage
Life cycle assessment	Eco-design	Life cycle assessment is a method of estimating environmental factors	Quantitative	Problem definition, conceptual design, and detailed design
Quality function deployment [12]	Eco-design	QFD is applied during the design phase to ensure that the requirement from the customer or market accuracy	Qualitative	Product planning Product development Process planning Production planning
ECQFD [13]	Partial-SPD	ECQFD is applied for translating voice of customer to engineering characteristics	Quantitative	Entire design stage
ECQFD and LCA-based method	Partial-SPD	Elements such as voice of customer, engineering metric, etc., are included	Quantitative	Entire design stage
LCA integrated with the Monte Carlo simulation	Partial-SPD	It integrates environmental, economic, and technical aspects and LCA is used to evaluate environmental performance	Quantitative	Conceptual design
Integrated ECQFD, TRIZ AND AHP	Partial-SPD	Integrate ECQFD, TRIZ, and AHP for sustainable product design	Quantitative	Conceptual design Problem definition
Integrated QFDE	Partial-SPD	Integrated QFDE with fuzzy decision making	Quantitative	All design
Fuzzy eco-QFD [14]	Partial-SPD	Integrate QFD with LCA in product design process and use fuzzy method to translate customer need into technical attribute	Quantitative	System-level design

5 Future Research Directions

We have seen that previously developed product design tool for EVs is mainly related to eco-design. This paper would reduce the effort of researchers in selecting the tools for sustainable product design of electric vehicles. The selection of the tool is very important to achieve the best result. This study would also improve the understanding of the tool in the field of electric vehicles. The table includes the application of most

Table 3 Categorization of sustainable product design tool for electric vehicle

Tool	Life cycle phases	Description	Analysis	Design stage
Life cycle sustainability assessment [15]	Raw material extraction, design and manufacturing, transformation, use, end of life	It gives triple bottom line of sustainability. Its gives relation between life cycle analysis, life cycle cost, and cost–benefit analysis	Quantitative	Detailed design
Fuzzy QFD [11]	Design and manufacturing, end of life	It is two-phase method and it depends on sustainability requirement and design consideration and useful for early-stage design and problem definition	Quantitative	Problem definition, conceptual design
Sustainable platform for product family design [16]	Material extraction, design and manufacturing, transportation, end of life	Measures risk value to calculate the robust component for redesign	Quantitative	Conceptual design
Fuzzy sustainability evaluation method [17]	Material extraction, design and manufacturing, transportation, use, end of life	Method depends upon three-dimensional stage of sustainability during detailed design	Quantitative	Problem definition

of the tools in the field of automobile and industrial areas. To achieve the best result in the sustainable product design of electric vehicles, the selection of the appropriate tools is very important.

There is much research done on sustainable product design of automobile vehicle parts but the sustainable design of EVs with different tool considerations will give future research direction. Sustainability analysis of EV parts like traction motor, battery, and motor controller with a different integration of tool is also a gap that may be performed for future research. Parameters like the voice of customers (VOC) and engineering metrics (EM) can be modified as per the product and process can be taken for the study. Identifying the best SPD tool in place of the partial-SPD tool can be taken because most of the researchers have applied the partial-SPD tool. Previously, ECQFD was used for the sustainable design of EVs but, in the future, the study can be focused on parts of EVs to improve sustainability. Future research may also base on the development of electric vehicle parts in CAD software with the integration of three aspects of sustainability like the environmental, economic, and social aspects [18]. More research is focused on the social aspect in comparison to environmental and economic.

6 Conclusion

Sustainable product design is a prevalent field of research area now a day. This chapter presents the literature and recent development on sustainable design and tools used for electric vehicles. This chapter provides better knowledge for researchers and practitioners working in this field by providing a systematic literature review of the published article in the renowned journal in the area of sustainable product design of EVs. It gives depth knowledge on sustainable design of EVs and sustainable design tools and techniques used to achieve sustainability. Previous research on the electric vehicle in the field of sustainable product design was based on the eco-design tool and partial-SPD tool. So, there was the need to apply recently developed SPD tools in electric vehicles to attain triple bottom line of sustainability. In this chapter, tools are categorized by different pillars of sustainability. Some authors have considered only two pillars of sustainability, viz. environment and economy. In this study, tools were categorized as partial-SPD. And some authors considered all three aspects of sustainability and in this study, such problem tools were categorized as SPD tools.

References

1. Ahmad S, Wong KY, Tseng ML, Wong WP (2018) Sustainable product design and development: a review of tools, applications and research prospects. Resour Conserv Recycl 132:49–61. https://doi.org/10.1016/j.resconrec.2018.01.020
2. Huang FH (2020) Understanding user acceptance of battery swapping service of sustainable transport: an empirical study of a battery swap station for electric scooters, Taiwan. Int J Sustain Transp 14(4):294–307. https://doi.org/10.1080/15568318.2018.1547464
3. Vinodh S, Rathod G (2010) Integration of ECQFD and LCA for sustainable product design. J Clean Prod 18(8):833–842. https://doi.org/10.1016/j.jclepro.2009.12.024
4. Vinodh S, Kamala V, Jayakrishna K (2014) Integration of ECQFD, TRIZ, and AHP for innovative and sustainable product development. Appl Math Model 38(11–12):2758–2770. https://doi.org/10.1016/j.apm.2013.10.057
5. Rathod G, Vinodhb S, Madhyasta UR (2011) Integration of ECQFD and LCA for enabling sustainable product design in an electric vehicle manufacturing organisation. Int J Sustain Eng 4(3):202–214. https://doi.org/10.1080/19397038.2010.547624
6. de Kwant C, Rahi AF, Laurenti R (2021) The role of product design in circular business models: an analysis of challenges and opportunities for electric vehicles and white goods. Sustain. Prod. Consum. 27:1728–1742. https://doi.org/10.1016/j.spc.2021.03.030
7. Tsang YP, Wong WC, Huang GQ, Wu CH, Kuo YH, Choy KL (2020) A fuzzy-based product life cycle prediction for sustainable development in the electric vehicle industry. Energies 13(15):3918
8. Kang X (2021) Combining rough set theory and support vector regression to the sustainable form design of hybrid electric vehicle. J Clean Prod 304:127137. https://doi.org/10.1016/j.jclepro.2021.127137
9. Song W, Niu Z, Zheng P (2021) Design concept evaluation of smart product-service systems considering sustainability: an integrated method. Comput Ind Eng 159. https://doi.org/10.1016/j.cie.2021.107485
10. Yang C, Tu J, Jiang Q (2020) The influential factors of consumers' sustainable consumption: a case on electric vehicles in China

11. Vinodh S, Manjunatheshwara KJ, Sundaram SK, Kirthivasan V (2017) Application of fuzzy quality function deployment for sustainable design of consumer electronics products: a case study. Clean Technol Environ Policy 19(4):1021–1030. https://doi.org/10.1007/s10098-016-1296-7
12. Zhang C, Huang H, Zhang L, Liu Z (2019) Semi-quantitative method for task planning in product eco-design. Int J Prod Res 57(8):2263–2280. https://doi.org/10.1080/00207543.2018.1514474
13. Bennett EB, Mingst A, Simon S, Graedel TE (2001) Improving the overall environmental performance of existing power generating facilities. IEEE Trans Energy Convers 16(3):234–238. https://doi.org/10.1109/60.937202
14. Yu S, Yang Q, Tao J, Xu X (2015) Incorporating quality function deployment with modularity for the end-of-life of a product family. J Clean Prod 87(1):423–430. https://doi.org/10.1016/j.jclepro.2014.10.037
15. Hoogmartens R, Van Passel S, Van Acker K, Dubois M (2014) Bridging the gap between LCA, LCC and CBA as sustainability assessment tools. Environ Impact Assess Rev 48:27–33. https://doi.org/10.1016/j.eiar.2014.05.001
16. Kim S, Moon SK (2017) Sustainable platform identification for product family design. J Clean Prod 143:567–581. https://doi.org/10.1016/j.jclepro.2016.12.073
17. Singh S, Olugu EU, Musa SN, Mahat AB (2018) Fuzzy-based sustainability evaluation method for manufacturing SMEs using balanced scorecard framework. J Intell Manuf 29(1):1–18. https://doi.org/10.1007/s10845-015-1081-1
18. Russo D, Rizzi C (2014) An ECO-DESIGN approach based on structural optimization in a CAD framework. Comput Aided Des Appl 11(5):579–588. https://doi.org/10.1080/16864360.2014.902691
19. Hasan MK, Mahmud M, Habib AA, Motakabber SMA, Islam S (2021) Review of electric vehicle energy storage and management system: standards, issues, and challenges. J Energy Storage 41:102940
20. Verma S, Dwivedi G, Verma P (2021) Life cycle assessment of electric vehicles in comparison to combustion engine vehicles: a review. Mater Today: Proc
21. Asghar R, Ullah K, Ullah Z, Waseem A, Ali N, Zeb K (2021) Assessment of the performance and shortcomings of common electric vehicle battery technologies. In: 2021 international conference on electrical, communication, and computer engineering (ICECCE). IEEE pp 1–6
22. Ramya KC, Ramani JG, Sridevi A, Rai RS, Shirley D (2022) Analysis of the different types of electric motors used in electric vehicles. In: E-mobility. Springer, Cham, pp 43–57
23. Mohammed SAQ, Jung JW (2021) A Comprehensive state-of-the-art review of wired/wireless charging technologies for battery electric vehicles: Classification/common topologies/future research issues. IEEE Access
24. Kong L, Wang L, Zhu J, Bian J, Xia W, Zhao R, Lin H, Zhao Y (2021) Configuring solid-state batteries to power electric vehicles: a deliberation on technology, chemistry and energy. Chem Commun 57(94):12587–12594
25. Tank I, Vignesh S, Bhateshvar YK, Vora KC (2021) Vehicle to grid (V2G): booster for EV adaptation in India. In: 2021 IEEE 2nd international conference on smart technologies for power, energy and control (STPEC). IEEE, pp 1–6
26. Kuvalekar SA, Mohanrajan SR (2021) PMSM torque ripple reduction in electric vehicle using neural network. In: 2021 IEEE international power and renewable energy conference (IPRECON). IEEE, pp 1–6
27. Zhao J, Xi X, Na Q, Wang S, Kadry SN, Kumar PM (2021) The technological innovation of hybrid and plug-in electric vehicles for environment carbon pollution control. Environ Impact Assess Rev 86:106506
28. Luo Y, Wu Y, Li B, Qu J, Feng SP, Chu PK (2021) Optimization and cutting-edge design of fuel-cell hybrid electric vehicles. Int J Energy Res 45(13):18392–18423
29. Gnann T, Stephens TS, Lin Z, Plötz P, Liu C, Brokate J (2018) What drives the market for plug-in electric vehicles?-A review of international PEV market diffusion models. Renew Sustain Energy Rev 93:158–164

30. El Bassam N (2021) Energy storage, smart grids, and electric vehicles. In: Distributed renewable energies for off-grid communities. Elsevier, pp 263–295
31. Shi D, Liu K, Wang Y, Liu R, Li S, Sun Y (2021) Adaptive energy control strategy of PHEV based on the Pontryagin's minimum principle algorithm. Adv Mech Eng 13(8):16878140211035598
32. Carignano M, Roda V, Costa-Castelló R, Valiño L, Lozano A, Barreras F (2019) Assessment of energy management in a fuel cell/battery hybrid vehicle. IEEE Access 7:16110–16122
33. Ferraren-De Cagalitan DDT, Abundo MLS (2021) A review of biohydrogen production technology for application towards hydrogen fuel cells. Renew Sustain Energy Rev 151:111413
34. Li M, Bai Y, Zhang C, Song Y, Jiang S, Grouset D, Zhang M (2019) Review on the research of hydrogen storage system fast refueling in fuel cell vehicle. Int J Hydrog Energy 44(21):10677–10693
35. Obaideen K, AlMallahi MN, Alami AH, Ramadan M, Abdelkareem MA, Shehata N, Olabi AG (2021) On the contribution of solar energy to sustainable developments goals: case study on Mohammed bin Rashid Al Maktoum Solar Park. Int J Thermofluids 12:100123
36. Devanathan S, Ramanujan D, Bernstein WZ, Zhao F, Ramani K (2010) Integration of sustainability into early design through the function impact matrix
37. Malhotra A, Rajak S, Jha SK (2019) An eco-innovative green design method by QFD and TRIZ tools-A case study of brass-ware manufacturing. Pertan J Sci Technol 27:2109–2121

Sustainability Assessment of Organizations Based on the Orientations of Product Sustainability

Jayakrishna Kandasamy, Barathi Venkatesh, and Aravind Raj Sakthivel

Abstract The main goal of this research paper is to develop sustainability assessment model for a manufacturing organization by considering manufacturing material, product design, and manufacturing methodology changes. Matrix Evaluation Method (MEM) is used as an assessment tool for measuring the sustainability level. The changes in three orientations are achieved by implementing the best alternatives. The best alternative for material, design, and manufacturing process was determined using Fuzzy VIsekriterijumska optimizacija iKOmpromisno Resenje (VIKOR), Environmentally Conscious Quality Function Deployment (ECQFD), and Life Cycle Analysis (LCA), respectively. The sustainability level, after performing the study has also been done. The model considers the changes in three perspectives such as material, product design, and manufacturing process for a product. This study showed that alternate material, product design, and manufacturing process for the instrument panel are, respectively, polypropylene, product design modifications in eco-friendly aspects, and vacuum forming as a replacement for injection molding. The sustainability index of the present and proposed models are 7.6 and 9.15, respectively. The conduct of the study enabled the practitioners to systematically assess sustainability level of the alternatives and select the best alternative. An attempt has been made to integrate Matrix Evaluation Method and Multiple Criteria Decision-Making tool for assessing the sustainability level and selecting the best alternative material, design, and process.

Keywords Sustainability assessment · Sustainability concepts · Fuzzy VIKOR · ECQFD · LCA

J. Kandasamy (✉) · B. Venkatesh · A. R. Sakthivel
School of Mechanical Engineering, VIT University, Vellore, Tamil Nadu, India
e-mail: mail2jaikrish@gmail.com

J. Kandasamy et al. (eds.), *Progress in Sustainable Manufacturing*, Management and Industrial Engineering, https://doi.org/10.1007/978-981-99-0201-9_4

Nomenclature

S_i	Utility index
R_i	Regret index
Q_i	VIKOR index
f_i^*	Best value
f_i^-	Worst value
vr_i	Improvement rate
AG	Aggregated value
DM	Decision-Maker
CR	Customer Ratings
EM	Engineering Metrics

1 Introduction

Sustainable development is a development that meets the needs of the present without compromising the ability of future generations to meet their own needs [2]. Sustainability assessment is a tool that can help decision-makers and policymakers decide what actions they should take and not take in an attempt to make society more sustainable [7]. Sustainability is also defined as the intersection of people, environment, and profit. Sustainability assessment is a tool that can help decision-makers and policymakers decide what actions they should take and not take in an attempt to make society more sustainable. In the present study, the economic and environmental dimensions of sustainable development are given importance; although social dimension of sustainability is also being concentrated by the case organization. The product life cycle in a manufacturing organization mostly depends on three factors such as material, design, and manufacturing process. The sustainability assessment was carried out by considering these three factors. In this method, the first step is to get the weights for the three factors from the decision-makers of the organization. One of the factors in sustainable development is material usage, i.e., resources utilization should be less and it should be recoverable. Decision made during the product design plays a major part in the resource consumption and waste generation and sustainability in manufacturing could be achieved through the selection of the best manufacturing process.

2 Literature Review

A literature survey was carried out extensively on sustainability assessment approaches carried out using MCDM techniques. The following subsections discuss the available literature.

2.1 Sustainability

Yang et al. [37] described the term sustainability assessment in a clear way by considering the philosophies that underscore various approaches available for sustainability and discussed the origin of sustainability assessment and existing methods for assessing sustainability. Mebratu [18] discussed the definition of sustainable development and pointed out the misconception regarding some of the definitions. Székely and Knrisch [29] explained various tools for sustainability evaluation such as sustainability indices, award schemes, accountability, and internal and external communication tools. These tools are majorly used by 20 German companies. The authors measured the sustainability performance of the case organization by considering three aspects of sustainability such as the economy, environment, and society. Vinodh [32] used multi-grade fuzzy approach for assessing sustainability in a manufacturing organization from the triple bottom line. Industrial implications of the case study are also explained. Labuschagne et al. [16] reviewed the sustainability methods such as global reporting initiative and presented the criteria considered under social, economic, and environmental dimensions for the assessment of sustainability. Becker [3] presented a conceptual framework for sustainability assessment of a land use system by considering four indicators such as economic, environmental, social, and composite and described the scope of different assessment tools using time and space matrix. Tseng [30] measured sustainable production indicators using a novel approach for a printed circuit board manufacturing sector and is composed of fuzzy set theory and a hierarchical structural model and they conducted a survey in two stages, in the first stage the researchers were asked to rate the ambiguity and clarity based on the outcome again they modified the instrument and in the second step, they repeated the same steps in order to have an accurate result. Musango and Brent [19] proposed a framework that incorporates a technology assessment approach, namely system approach to technological sustainability assessment (SATSA) that integrates three key elements: technology development, sustainable development, and dynamic systems approach. Ren et al. [27] proposed to rank the prior arrangement of biomass-based technologies for hydrogen production by using a sustainability assessment method.

2.2 VIKOR

Prato [23] explained the disadvantages of stochastic and nonstochastic approaches used for the sustainability assessment and to overcome those disadvantages, they introduced fuzzy logic methods to assess and rank the management alternatives. Phillis and Andriantiatsaholiniaina [24] developed a model called Sustainability Assessment by Fuzzy Evaluation (SAFE), which combines both human and ecological inputs and treated them together with the aid of fuzzy logic to provide an overall measure. Hîncu [11] modeled the urban sustainable development using fuzzy sets by implementing some new steps in the SAFE and used trapezoidal fuzzy numbers for their evaluation. Girubha and Vinodh [10] developed a framework to select the best material for the particular application. They concluded that polypropylene was the alternative material for the instrument panel by using MCDM and VIKOR assessment methosd. Devi [8] discussed VIKOR method is extended to the intuitionistic fuzzy environment and to solve multiple-criteria decision-making problems. For application and verification, this study presents a robot selection problem for material handling task.

2.3 ECQFD

Heijungs [12] developed a framework to incorporate different models for environmental analysis including the social and economic aspects and they analyzed the cost expenditure from private to societal. They concluded that the life cycle sustainability assessment is the sum of life cycle costing, social life cycle analysis, and life cycle analysis. Bereketli and Genevois [5] reviewed to identify the improved strategies in sustainable development by integrating a multi-aspect QFD and Environment (QFDE). Pope et al. [22] compared two types of approaches used for the assessment of sustainability such as triple bottom line approach that includes Environmental Impact Assessment (EIA)-driven integrated assessment, objectives-led integrated assessment, and assessment for sustainability and principle-based approaches and concluded that the appropriate assessment can be done by principle-based approaches. Wu and Ho [35] used to integrate green quality function deployment and fuzzy theory for assessing sustainability in green design. Green mobile phone design suggestions of the case study are also explained. Waheed et al. [34] discussed different approaches to select a framework for the effective sustainability assessment and mainly discussed the advantages and disadvantages related to the linkage-based frameworks and the driving force-state-exposure-effect-action (DPSEEA) framework. Mir et al. [21] proposed the MCDM methods to evaluate the finest municipal solid waste management method for matching and ranking the scenarios. They concluded that the anaerobic digestion and a sanitary landfill with Electricity Production (EP) methods are the favorite choices for MSWmanagement. Romli et al. [28] proposed a framework that integrates eco-design decision-making (IEDM) methodology it is formed by life cycle assessment, an eco-design process (Eco-Process) model, and an enhanced eco-design quality function deployment process.

2.4 LCA

Xu et al. [36] proposed to improve the service quality of the customers by assessing the service performance of EVS programs. They conducted sensitivity analysis and a comparative analysis of different methods and conclude that the proposed method is capable to assess service performance of EVS programs. Ameknassi et al. [1] discussed the implementation of DfE activities to offer an integrated approach to support designers efficiently. Little differences in the PDP structure are made to express the organizational behaviors of the approach to providing modernized specific detail, and to make an order of the system in the communication among executors and proponents of DfE. Paju et al. [25] proposed a value stream mapping, life cycle analysis, and discrete event-based assessment tool, termed Sustainable Manufacturing Mapping (SMM). The main phases included in this work are goal definition, identification of the sustainability indicators, and modeling the current and future state maps. Zhao et al. [38] introduced technology assessment as a feasibility study tool for sustainability and to increase its application, included the instruments such as LCA, EIA, or environmental audit, and the interactions between them. Kalhor et al. [15] investigated the life cycle assessment to evaluate the chicken meat generation at slaughterhouse gate per mass-based functional unit and the environmental effect of broiler generation at farm gate in summer and winter periods. The outcomes of environmental problems of chicken meat generation in winter were higher than summer period. Isaksson and Garvare [14] proposed a framework model to assess the sustainability of cement manufacturing units and they listed out the effects of CO_2. They concluded that energy and material consumption was high in machine tools and suggested ways to overcome those effects. Tadić et al. [31] explained about the city logistics concept frame work selection which would be most suitable for different participants, stakeholders, and which would fulfill with attributes of the surroundings.

2.5 *Literature Gap*

Based on the literature review it is found that the researchers have focused on developing product-based sustainability model using multi-criteria decision-making techniques. But in this research, a new attempt has been made in computing the product sustainability index using fuzzy VIKOR, Environmentally Conscious Quality Function Deployment (ECQFD) and Life Cycle Analysis (LCA) across the three major orientations of product sustainability (i.e. material, product design and manufacturing cost).

3 Methodology

The case study was carried out in an automotive component manufacturing industry located in Bengaluru, India. An automotive component manufacturing was chosen because it was more susceptible to sustainable development and more benefits could be obtained if the research is properly conducted. The study was carried out on the instrument panel. The instrument panel is the dashboard and includes all the equipments such as speedometer which is situated directly ahead of the vehicle's driver. The industry has initiated steps to find out the best material, product design, and the manufacturing processes. Before implementing the changes, the organization was in need of assessing the sustainability level. Therefore in this study, the assessment is carried out by considering those factors by getting the inputs from the past analysis and decision-makers. The decision-makers involved all sections of the organization starting from the top chain to the last chain in order to get all the information required to carry out the required research such as Chief Executive Officer, Chief Technical Officer, Head of Departments and Site Supervisors. The decision-makers participated in the study possess wide experience regarding the operational culture of the organization. Inputs for the methodologies used in this study were gathered from the decision-makers. The inputs were provided by the decision-makers considering the sustainability characteristics prevailing in the organization as well as to improve the competitive position of the organization. The methodology followed during this study is shown in Fig. 1. The study begins with exploring the sustainability assessment model for incorporating the three sustainable manufacturing concepts. Matrix evaluation model was used in this case study. The model compares the sustainability level before and after implementing the alternatives for each concept. The alternate material is determined using Fuzzy VIKOR methodology. Then the product design modifications have been implemented using ECQFD and finally, the manufacturing process selection was based on LCA.

3.1 Sustainability Assessment Model

The assessment work was carried out with material, product design, and manufacturing process using matrix evaluation model. This project aims at achieving sustainability in material, design and manufacturing process levels. So the sustainability level needs to be measured before and after incorporating the alternatives for three orientations.

3.2 Formation of the Evaluation Matrix

The evaluation matrix was formulated by considering three main criteria and their sub criteria i.e., at two hierarchical levels. The model used for the evaluation was shown in Fig. 2. The alternatives for three concepts such as material, product design and manufacturing process were chosen by Fuzzy VIKOR [4, 26] ECQFD [6, 20]and LCA [13]. The sub criteria were chosen by taking these three aspects into account for evaluation using the metrics achieved in the above three methods. The weighting for each criterion was obtained from the decision-makers. They provided the weight by

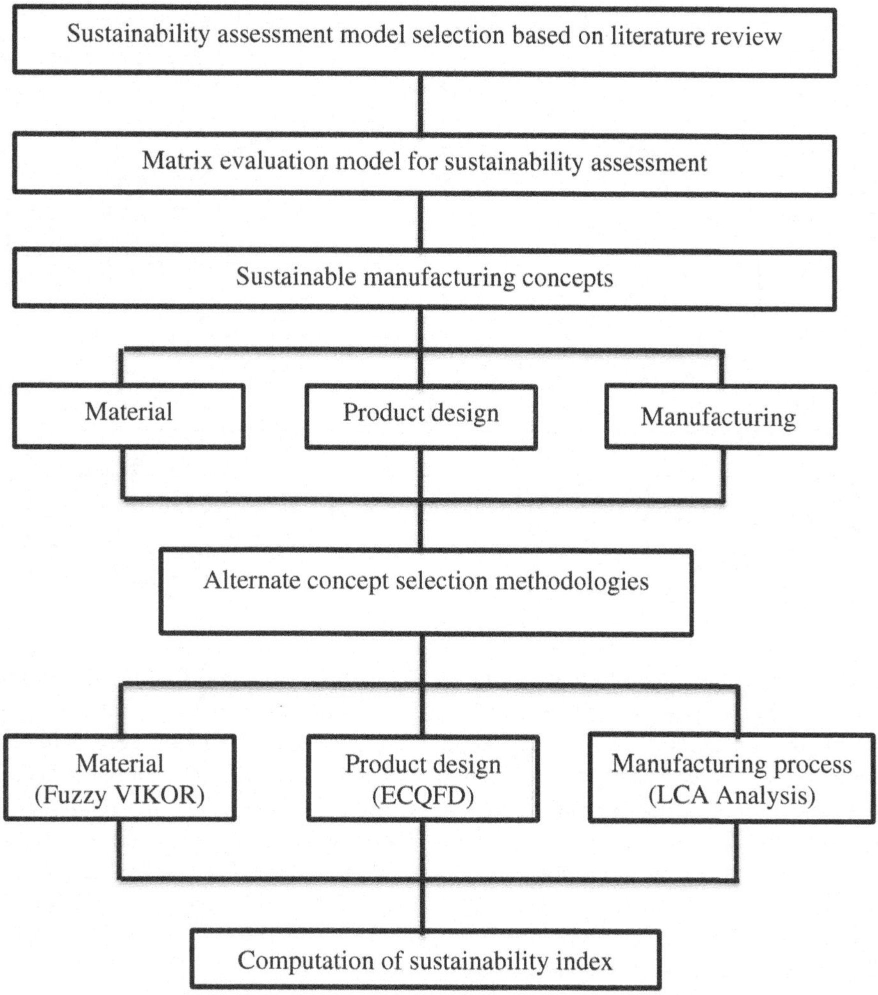

Fig. 1 Methodology for the study

considering the importance of the criteria with respect to manufacture of a sustainable product.

3.3 *Material Selection Using Fuzzy VIKOR*

The current material used by, the organization is ABS. To improve the sustainability level of any product, there exists a need to find an alternate material. Therefore, the

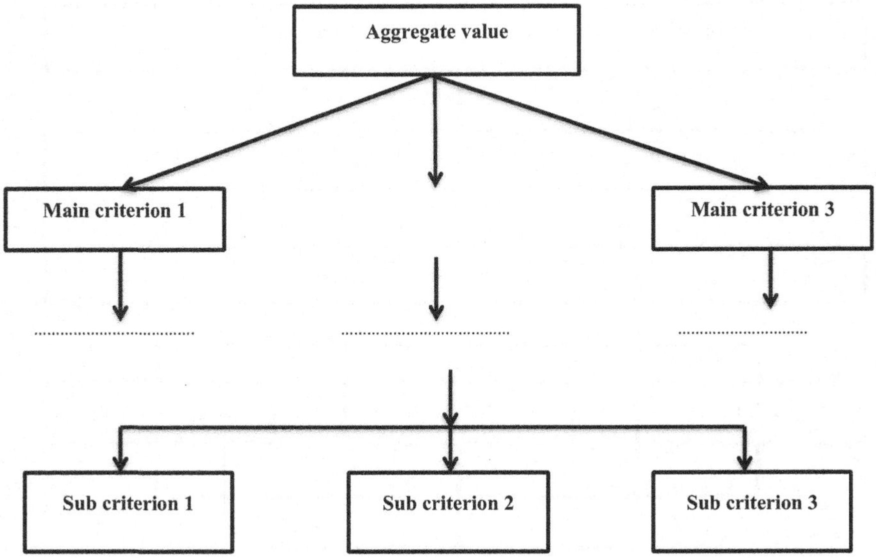

Fig. 2 Matrix evaluation model

decision-makers suggest using an efficient fuzzy-based MCDM tool for the material selection. This leads to the selection of fuzzy VIKOR for the material selection problem [4]. The selection is made among four materials including the current material. The inputs required for the selection of materials are acquired in the form of fuzzy numbers in order to avoid vagueness associated with inputs. Trapezoidal fuzzy numbers are used in this selection process. The linguistic variables and the corresponding fuzzy sets are as follows Very Poor (VP) – (0.0, 0.0, 0.1,0.2), Poor (P) – (0.1, 0.2, 0.2, 0.3), Medium Poor (MP) – (0.2, 0.3, 0.4, 0.5), Fair (F) – (0.4, 0.5, 0.5, 0.6), Medium Good (MG) – (0.5, 0.6, 0.7, 0.8), Good (G) – (0.7, 0.8, 0.8, 0.9), Very Good (VG) – (0.8, 0.9, 1.0, 1.0) [9].

By considering the functionalities of the component, it is evident that the following eight criteria have to be taken into account for the material selection. Temperature withstanding capability (C1), Capability to recycle (C2), Price (C3), Strength to Weight (C4), Thermal conductivity (C5), Tensile strength (C6), Stiffness (C7), Eco friendliness (C8) [4, 33].

The inputs have been collected in two forms, first one as weights of the criteria and the second one deals about the decision-makers view about the material with respect to each criterion. An excerpt of the weights of the criterion is shown in Table 1 and an excerpt of the ratings of criterion against each material provided by decision-maker 1 is shown in Table 2.

After getting the linguistic variables as inputs from the decision-makers, the inputs have to be aggregated into a single set of value. In order to get a crisp value, an average of the four numbers is used and the crisp values are shown in Table 3.

Table 1 Excerpt of the weights of the criteria

Decision maker / Criteria	DM 1	DM 2	DM 3	DM 4	DM 5
C3	G	G	MG	MG	G
C4	G	MG	VG	G	G
C5	G	VG	G	VG	G

Table 2 Excerpt of the ratings of the criteria against each material by decision maker 4

Criteria / Material	C1	C2	C3	C4	C5	C6	C7	C8
M1	FH	H	FH	H	FH	FH	H	FH
M2	FH	FH	H	H	H	H	FH	H
M3	VH	VH	H	H	VH	VH	H	H
M4	H	VH	H	H	H	VH	H	H

Table 3 Crisp values of the weights and ratings of the material

	C1	C2	C3	C4	C5	C6	C7	C8
W	0.77	0.77	1.79	0.77	2.13	1.76	2.11	1.76
M1	0.69	0.72	1.73	0.70	1.73	1.79	1.73	1.73
M2	0.65	0.61	1.68	0.66	1.68	1.63	1.57	1.60
M3	0.87	0.87	2.13	0.85	2.13	2.16	2.13	2.11
M4	0.80	0.87	2.13	0.80	2.13	2.13	2.16	1.84

The best and worst values among the material for each criterion are found and it is shown in Table 4. The calculation of utility, regret and VIKOR index is shown in Table 5 using Eqs. 1–3 [4].

$$S_i = \sum_{j=1}^{n} \frac{w_j^o\left(f_i^* - f_{ij}\right)}{\left(f_i^* - f_i^-\right)} \tag{1}$$

$$R_i = \max_i \left(\frac{w_j^o\left(f_i^* - f_{ij}\right)}{\left(f_i^* - f_i^-\right)} \right) \tag{2}$$

$$Q_i = \frac{v(S_i - S^*)}{S^- - S^*} + \frac{(1 - v)(R_i - R^*)}{R^- - R^*} \tag{3}$$

The ranking of material is shown in Table 6 and it shows the current material occupies the second position and the proposed material is selected as the best material. i.e., the current material being used is ABS, which occupies the second position;

Table 4 Best and worst values

f_i^*	0.87	0.87	2.13	0.85	2.13	2.16	2.16	2.11
f_i^-	0.65	0.61	1.68	0.66	1.68	1.63	1.57	1.6

Table 5 Calculation of utility, regret measure and VIKOR index

	M1	M2	M3	M4
S_i	9.2	11.9	0.11	1.4
R_i	1.8	2.1	0.11	0.93
Q_i	0.85	1.00	0.00	0.35

Table 6 Ranking of materials

	1	2	3	4
S_i	M3	M4	M1	M2
R_i	M3	M4	M1	M2
Q_i	M3	M4	M1	M2

whereas Polypropylene is selected as the best material, since it occupies the first position.

3.4 Design Modifications Using ECQFD

ECQFD can be applied to find out the environmentally friendlier modifications for a product [33]. The design modifications of the instrument panel are made using ECQFD. Currently, the case organization is giving more importance to four major aspects such as increased life, reduced weight, volume and number of parts. In order to increase the sustainability level, three new environmental aspects are taken and compared with the current attempt to modify the design form environment perspective. The three new environmental aspects include biodegradability, the toxicity of materials and increased hardness. The result shows that the new environmental perspectives could improve the sustainability than the present attempt. The methodology of ECQFD is explained below.

3.4.1 Environmental Voice of Customers (VOCs) and Environmental Engineering Metrics (EMs)

Reduced material usage, energy consumption; increased durability; easy to disassemble during the maintenance stage; harmless to the living environment of the users during manufacturing and use; and ease of dispersibility represents the Environmental VOCs of the case component. The technical characteristics of instrument panel assembly, such as reduced weight, volume, the number of parts, increased life, hardness or strength, biodegradability and toxicity of materials used represents the Environmental EMs of the case component.

3.4.2 Identification of Parameters for Product Design Improvement

This section describes the application of ECQFD for changes in product design and process sustainability.

ECQFD Phase I

Phase I explains the application of ECQFD to the instrument panel assembly. The normal raw score for each object is shown in Table 7. For example, EM Engineering Metrics such as "increased life", "reduced weight", "reduced volume" and "reduced number of parts" with normal raw scores of "0.137", "0.127", "0.131" and "0.140" respectively are reasonably important to satisfy the user requirements than the "increased hardness or strength", "biodegradability" and "toxicity".

ECQFD Phase II

Options I and II with respect to Phase II, as decided by the design engineers are shown in Tables 8 and 9. The two design options selected were based on the outcomes of phase I and II and incorporating the following combinations of components and EMs Engineering Metrics.

The correlating points between the target EM and components are shown in Tables 8 and 9. The improvement rate of each EM Engineering Metrics item 'mr_j' was obtained from Eq. 4, [33].

$$\mathrm{mr}_j = \frac{\sum_{k=1}^{K}\left(b_{j,k}c_{j,k}\right)}{\sum_{k=1}^{K}\left(b_{j,k}\right)} \quad (J = 1,\ 2,\ \ldots,\ J) \tag{4}$$

where

k is the index number of component
J is the index number of EM
$b_{j,\,k}$ is the relational strength between EM item j to component k
$c_{j,\,k}$ is the improvement rate of EM Engineering Metrics item j to component k and originally allowed to take the real number from 0.0 to 1.0

Table 7 ECQFD phase I

EM VOC	Customer weights	Increased hardness/ strength	Increased biodegradability	Reduced Toxicity of materials	Increased life	Reduced weight	Reduced volume	Reduced number of parts
Reliability	**5**	5	4	5	2	2	4	1
Durability	**2**	5	4	4	1	3	5	5
Less energy consumption	**2**	4	2	4	4	4	4	2
Ease of disassembly	**4**	3	3	4	4	4	4	5
Less material usage	**1**	3	2	3	3	3	3	4
Harmless to living environment	**5**	5	4	3	4	4	4	4
Aesthetics	**2**	4	5	2	4	3	3	3
Recycling of parts	**1**	4	5	5	4	5	4	4
Ease of disposal	**5**	4	5	4	5	3	1	5
Raw score		115	106	104	96	89	92	98
Relative weight		0.164	0.151	0.149	0.137	0.127	0.131	0.140

Table 8 ECQFD phase II for option I

Engineering metrics engineering metrics	Phase I relative weight	Part No																	Score	Improvement rate of engineering metrics
		1	2	3	4	5	6	7	8	9	10	11	12	13	14	15	16	17		
Hardness/strength	0.164	0	0	0	0	0	2	0	0	4	2	0	3	0	0	0	0	0	11.00	3.860
Biodegradability	0.151	0	0	2	0	5	0	0	2	0	0	3	0	0	0	0	4	0	16.00	3.711
Toxicity of materials	0.149	4	5	0	0	3	0	0	0	0	3	0	0	5	0	0	1	0	21.00	3.120
Increased life	0.137	0	0	0	0	0	0	0	0	0	0	0	0	0	0	0	0	0	0.00	0.000
Reduced weight	0.127	0	0	0	0	0	0	0	0	0	0	0	0	0	0	0	0	0	0.00	0.000
Reduced volume	0.131	0	0	0	0	0	0	0	0	0	0	0	0	0	0	0	0	0	0.00	0.000
Reduced number of parts	0.140	0	0	0	0	0	0	0	0	0	0	0	0	0	0	0	0	0	0.00	0.000

Table 9 ECQFD phase II for option II

Engineering metrics engineering metrics	Phase I relative weight	Part No																	Score	Improvement rate of engineering metrics
		1	2	3	4	5	6	7	8	9	10	11	12	13	14	15	16	17		
Hardness/strength	0.164	0	0	0	0	0	0	0	0	0	0	0	0	0	0	0	0	0	0.00	0.000
Biodegradability	0.151	0	0	0	0	0	0	0	0	0	0	0	0	0	0	0	0	0	0.00	0.000
Toxicity of materials	0.149	0	0	0	0	0	0	0	0	0	0	0	0	0	0	0	0	0	0.00	0.000
Increased life	0.137	2	4	2	0	1	2	0	0	3	3	0	3	0	5	0	0	2	27.00	3.703
Reduced weight	0.127	0	0	0	3	0	0	0	3	0	0	2	0	0	1	0	2	0	11.00	1.399
Reduced volume	0.131	0	3	2	2	0	3	0	0	2	2	0	2	0	0	3	0	3	22.00	2.891
Reduced number of parts	0.140	0	0	0	0	0	0	2	0	0	0	2	0	0	0	0	0	0	4.00	0.560

Hence, for simplicity, $c_{j,k}$ can take the binary numbers:
$c_{j,k} = 1$ (improvement possible)
$c_{j,k} = 0$ (improvement impossible).

ECQFD Phase III

The improvement rate for each environmental VOC 'vr_i' is obtained from the following Eq. 5, [9].

$$\mathrm{vr}_i = \frac{\sum_{j=1}^{J}\left(\mathrm{mr}_j a_{i,j}\right)}{\sum_{j=1}^{J}\left(a_{i,j}\right)} \quad (i = 1, 2, \ldots, J) \tag{5}$$

where J is the index number of EM item
I is the index number of a VOC item
$a_{i,j}$ is the relational strength between VOC item I and EM Engineering Metrics item j.

Evaluation of Design for Environment Options

The design improvements such as increasing bio degradability, reducing the toxicity of materials and improving the hardness and strength can be modified by choosing the appropriate material for the instrument panel. Here in this context, these changes can be fulfilled by the usage of polypropylene.

The improvement effect of the VOCs and their respective weight was calculated based on environmental perspective through phases II and III. From this case study, the scores 14.568 and 8.561 were obtained for options I and II respectively and it was confirmed that option I is higher compared to option II. This indicates that three new concepts can be considered in order to increase the sustainability level.

3.5 Process Sustainability Using Life Cycle Assessment

For comparing the manufacturing processes from the sustainability perspective, an analysis was made using GaBi, a Life cycle Assessment (LCA) software package. It mainly deals with the LCA of the two processes such as vacuum forming and injection molding. Currently, the organization is using injection molding for manufacturing the instrument panel. Therefore in this study, an attempt was made to compare vacuum forming and injection molding processes.

An attempt was made to study the sustainability of a process using GaBi. Two life cycles were modeled for the same product. The parameters and units of measurement are provided in Tables 10 and 11.

3.5.1 Inferences from LCA

On comparing vacuum forming with injection molding the following influences are being derived.

- The energy required for injection molding process was high when compared to vacuum forming and it is shown in Table 12.
- Emissions which affect the atmosphere that are caused by injection molding are high when compared with vacuum forming process and it is shown in Table 13.

The overall sustainability of vacuum forming process is better than the injection molding process.

4 Sustainability Index Calculations

The three criteria are inter-related so they are included as a sub criterion for the three factors. The quantifiable details are entered as the sub criterion values and the normalized matrix is formed.

The comparison between the alternative for three concepts are taken from the above three methodologies such as fuzzy VIKOR, ECQFD, and LCA. The best alternate and the present technology have been taken for the comparison purpose and this forms the metrics for the evaluation matrix and is shown in Table 14.

The aggregated value is calculated using the equation Eq. 6, [37].

$$\mathrm{AG} = \frac{\sum_{i=1}^{I}\left\{w_i\left(\sum_{j=1}^{J}\frac{v_{ij}}{\sum \max(v_{ij}{}^{(k)})}\right)\right\}}{\sum_{i=1}^{I} w_i} \tag{6}$$

where

v_{ij} refers to the normalized value obtained for every sub criterion
W_i is the weight obtained by the MCDM method for every main criterion
I, J denotes the maximum number of criterion and sub- criterion.

The aggregated values of the three criteria are presented in Table 15. The aggregated value for beneficial criteria is calculated by finding out the sum of present and proposed values in Table 14, and then by dividing the respective values with the sum, yields the corresponding value in Table 15. The aggregated value for rest of the criteria is calculated by finding out the inverse of the proposed and present values.

Table 10 ECQFD phase III for option I

EM engineering metrics VOC	Customer weights	Increased Hardness/ strength	Increased Biodegradability	Reduced Toxicity of materials	Increased life	Reduced weight	Reduced Volume	Reduced number of parts	Improvement rate of CR	Improvement effect of CR
Reliability	**5**	5	4	5	2	2	4	1	0.433	2.163
Durability	**2**	5	4	4	1	3	5	5	0.863	1.727
Less energy consumption	**2**	4	2	4	4	4	4	2	0.736	1.473
Ease of disassembly	**4**	3	3	4	4	4	4	5	0.326	1.303
Less material usage	**1**	3	2	3	3	3	3	4	1.351	1.351
Harmless to living environment	**5**	5	4	3	4	4	4	4	0.311	1.554
Aesthetics	**2**	4	5	2	4	3	3	3	0.838	1.676
Recycling of parts	**1**	4	5	5	4	5	4	4	1.600	1.600
Ease of disposal	**5**	4	5	4	5	3	1	5	0.344	1.721
Raw score		3.860	3.711	3.120	0.000	0.000	0.000	0.000	6.802	14.568
Amount									**6.802**	**14.568**

Table 11 ECQFD phase III for option II

EM engineering metrics VOC	Customer weights	Increased Hardness/ strength	Increased Biodegradability	Reduced Toxicity of materials	Increased life	Reduced weight	Reduced Volume	Reduced number of parts	Improvement rate of CR	Improvement effect of CR
Reliability	**5**	5	4	5	2	2	4	1	0.433	2.163
Durability	**2**	5	4	4	1	3	5	5	0.863	1.727
Less energy consumption	**2**	4	2	4	4	4	4	2	0.736	1.473
Ease of disassembly	**4**	3	3	4	4	4	4	5	0.326	1.303
Less material usage	**1**	3	2	3	3	3	3	4	1.351	1.351
Harmless to living environment	**5**	5	4	3	4	4	4	4	0.311	1.554
Aesthetics	**2**	4	5	2	4	3	3	3	0.838	1.676
Recycling of parts	**1**	4	5	5	4	5	4	4	1.600	1.600
Ease of disposal	**5**	4	5	4	5	3	1	5	0.344	1.721
Raw score		3.860	3.711	3.120	0.000	0.000	0.000	0.000	6.802	14.568
Amount									**6.802**	**14.568**

Table 12 Energy resource consumption

Energy resources	Injection molding	Vacuum forming
Non renewable energy resources	100.35	76.362
Renewable energy resources	0.0004	0.00015
Non renewable elements	7.01	4.04
Non renewable resources	1268.714	1151.581
Renewable resources	1993.23	1928.375

Table 13 Comparison of emissions

Emissions	Injection molding	Vacuum forming
Emissions to air	360.36	100.265
Emissions to fresh water	2.369	1.023
Emissions to sea water	0.356	0.956
Emissions to industrial soil	9.95E-05	3.23E-05
Heavy metals to air	0.000125	0.000123
Inorganic emissions to air	313.077	260.53
Organic emissions to air	0.0145	0.0093
Other emissions to air	800.36	432.23
Particles to air	0.0198	0.009

The final assessment of the alternate concept from the evaluation matrix is the aggregation of the weighted sum of all the sub-criteria evaluations with the non-linear normalizations. The result obtained from the computations is shown in Tables 16, 17 and 18 and it shows that the sustainability index is 7.65 for the present concepts and for the proposed alternate is 9.15. This shows that there is an improvement in sustainability by substituting the alternatives for three concepts.

5 Practical Implications

The conduct of the study will enable the industry decision-makers to understand the importance of MEM for sustainability assessment. Effective training needs to be provided on the selection of appropriate tools/techniques for three major sustainability orientations. In the present study, the decision-makers gained knowledge on the practical propensity of deploying MCDM for selecting but material, ECQFD application for identifying best design options from environment viewpoint and LCA deployment for identifying potential components from a sustainability viewpoint. Also, the study enabled the identification of appropriate criteria for three sustainability perspectives as well as the computation of sustainability index.

Table 14 Metrics of the three orientations

Orientations	Criteria	Present concept	Proposed concept
Material	*Criteria*	*ABS*	*PP*
	Max temp limit	0.8	0.87
	Recyclability	0.87	0.87
	Elongation	2.13	2.13
	Weight	0.8	0.85
	Thermal conductivity	2.13	2.13
	Tensile strengt	2.13	2.16
	Cost	2.16	2.13
	Toxicity level	1.84	2.11
Product design	*Criteria*	*Option II*	*Option I*
	Reliability	0.97	1.79
	Durability	0.93	1.44
	Less energy consumption	1.38	1.22
	Ease of disassembly	1.29	1.07
	Less material usage	1.25	1.12
	Harmless to living environment	1.22	1.31
	Asthetics	1.22	1.42
	Recycling of parts	1.15	1.31
	Ease of disposal	1.05	1.43
Manufacturing process	*Criteria*	*Injection molding*	*Vacuum forming*
	Non renewable energy resources	100.350	76.362
	Renewable energy resources	0.0004	0.00015
	Non renewable elements	7.010	4.040
	Non renewable resources	1268.714	1151.581
	Renewable resources	1993.230	1928.375
	Emissions to air	360.360	100.265
	Emissions to fresh water	2.369	1.023
	Emissions to sea water	0.356	0.956
	Emissions to industrial soil	0.00010	0.0003
	Heavy metals to air	0.00013	0.00012
	Inorganic emissions to air	313.077	260.530
	Organic emissions to air	0.015	0.009
	Other emissions to air	800.360	432.230
	Particles to air	0.020	0.009

Table 15 Evaluation table with weights

Main criteria	Sub criteria	Value of present concept $v_{ij}^{(1)}$	Value of proposed alternate $v_{ij}^{(2)}$	Weight
Material	Max temp limit	0.4790	0.5210	1
	Recyclability	0.5000	0.5000	
	Elongation	0.5000	0.5000	
	Weight	0.4855	0.5145	
	Thermal conductivity	0.5000	0.5000	
	Tensile strength	0.4965	0.5035	
	Cost	0.5035	0.4965	
	Toxicity level	0.4658	0.5342	
Product design	Reliability	0.3517	0.6483	2
	Durability	0.3925	0.6075	
	Less energy consumption	0.5312	0.4688	
	Ease of disassembly	0.5471	0.4529	
	Less material usage	0.5278	0.4722	
	Harmless to living environment	0.4824	0.5176	
	asthetics	0.4630	0.5370	
	Recycling of parts	0.4664	0.5336	
	Ease of disposal	0.4245	0.5755	
Manufacturing process	Non renewable energy resources	0.4321	0.5679	3
	Renewable energy resources	0.2727	0.7273	
	Non renewable elements	0.3656	0.6344	
	Non renewable resources	0.4758	0.5242	
	Renewable resources	0.4917	0.5083	
	Emissions to air	0.2177	0.7823	
	Emissions to fresh water	0.3016	0.6984	
	Emissions to sea water	0.7287	0.2713	
	Emissions to industrial soil	0.2451	0.7549	

(continued)

Table 15 (continued)

Main criteria	Sub criteria	Value of present concept $v_{ij}^{(1)}$	Value of proposed alternate $v_{ij}^{(2)}$	Weight
	Heavy metals to air	0.4960	0.5040	
	Inorganic emissions to air	0.4542	0.5458	
	Organic emissions to air	0.3908	0.6092	
	Other emissions to air	0.3507	0.6493	
	Particles to air	0.3125	0.6875	

Table 16 Calculation of sum and maximum values

Criteria	Sum of present	Sum of proposed	Max of present	Max of proposed
Material	3.9303	4.0697	0.5035	0.5342
Product design	4.1866	4.8134	0.5471	0.6483
Manufacturing process	5.535077	8.464923	0.728659	0.782328

Table 17 Multiplied present and proposed values

Criteria	Sum of present/max of present	Sum of proposed/max of proposed	Weight	Multiplied value of present value with weight	Multiplied value of proposed value with weight
Material	7.806016	7.618382	1	7.806016	7.618382
Product design	7.652469	7.424533	2	15.30494	14.84907
Manufacturing process	7.596256	10.82017	3	22.78877	32.4605
Sum			6	45.62188	54.94669

Table 18 Calculation of sustainability index

Sum of final present/sum of weights	Sum of proposed/sum of weights
7.603646	9.157782

6 Conclusions

In order to achieve sustainability on the whole for a product, there is a necessity to integrate three main orientations such as material, design, and manufacturing process of a product. The sustainability assessment studies reported in the literature do not focus on the three major orientations comprehensively to the best knowledge of the authors.

Therefore, there is a need to select the best alternative in every concept. Various challenges were faced while performing this research which includes setting the goal while giving due consideration to the boundary conditions while performing LCA. Another challenge was making the concerned authority feel the need for sustainable development and then implementing the required changes. Sustainability assessment tool namely, matrix evaluation method is used to assess the sustainability level of the present and proposed alternatives for three sustainability manufacturing concepts. Fuzzy VIKOR was used as the material selection methodology it was proposed that Polypropylene could be the best alternative for ABS. ECQFD was used to find out the modifications in the design of the instrument panel and it was suggested to improve the Hardness/strength, increase biodegradability, and to reduce the toxicity of materials. LCA was carried out, in order to find out the best alternate manufacturing process for producing instrument panel and it was suggested that the vacuum forming could be the best alternative for injection molding. By taking these the new alternatives, the sustainability level was assessed by including the weights for the three orientations and it shows that the sustainability level before and after implementing the changes is 7.65 and 9.165, respectively. The assessment shows that the sustainability is in the range of 7–8 for the present orientations and for the proposed alternates for the orientations, it is found to be in the range of 9–10. This shows that there is an improvement of 20% in attaining sustainability. Therefore, the sustainability level of the product is increased by incorporating the changes in three major sustainability orientations. Making the sustainability index aids in benchmarking the industry to various other similar industries which would result in further improvement of the model.

In the present study, matrix evaluation model was used as the technique for sustainability assessment encompassed with the usage of appropriate technique in three sustainability orientations. In future, similar case studies could be conducted using the proposed approach for enhancing the cumulative effectiveness.

References

1. Ameknassi L, Ait-Kadi D, Keivanpour S (2016) Incorporating design for environment into product development process: an integrated approach. IFAC-Pap Online 49(12):1460–1465
2. Brundtland GH (1987) Report of the World Commission on environment and development: our common future. United Nations
3. Becker B (1997) Sustainability assessment: a review of values, concepts, and methodological approaches. The World Bank
4. Bahraminasab M, Jahan A (2011) Material selection for femoral component of total knee replacement using comprehensive VIKOR. Mater Des 32(8):4471–4477
5. Bereketli I, Genevois ME (2013) An integrated QFDE approach for identifying improvement strategies in sustainable product development. J Clean Prod 54:188–198
6. Cristofari M, Deshmukh A, Wang B (1996) Green quality function deployment. In: Proceedings of the 4th international conference on environmentally conscious design and manufacturing, pp 297–304
7. Devuyst D, Hens L, De Lannoy W (eds) (2001) How green is the city?: sustainability assessment and the management of urban environments. Columbia University Press

8. Devi K (2011) Extension of VIKOR method in intuitionistic fuzzy environment for robot selection. Expert Syst Appl 38(11):14163–14168
9. De Carvalho JF (2011) Measuring economic performance, social progress and sustainability using an index. Renew Sustain Energy Rev 15(2):1073–1079
10. Girubha RJ, Vinodh S (2012) Application of fuzzy VIKOR and environmental impact analysis for material selection of an automotive component. Mater Des 37:478–486
11. Hîncu D (2011) Modelling the urban sustainable development by using fuzzy sets. Theor Empir Res Urban Manag 6(2):88–103
12. Heijungs R (2010) Ecodesign—Carbon footprint—Life cycle assessment—Life cycle sustainability analysis. A flexible framework for a continuum of tools. Sci J Riga Tech Univ. Environ Clim Technol 4(1):42–46
13. Heijungs R, Huppes G, Guinée JB (2010) Life cycle assessment and sustainability analysis of products, materials and technologies. Toward a scientific framework for sustainability life cycle analysis. Polym Degrad Stab 95(3):422–428
14. Isaksson R, Garvare R (2003) Measuring sustainable development using process models. Manag Audit J 18(8):649–656
15. Kalhor T, Rajabipour A, Akram A, Sharifi M (2016) Environmental impact assessment of chicken meat production using life cycle assessment. Inf Process Agric 3(4):262–271
16. Labuschagne C, Brent AC, Van Erck RP (2005) Assessing the sustainability performances of industries. J Clean Prod 13(4):373–385
17. Ludwig B (1997) The concept of technology assessment-an entire process to sustainable development. Sustain Dev 5(3):111–117
18. Mebratu D (1998) Sustainability and sustainable development: historical and conceptual review. Environ Impact Assess Rev 18(6):493–520
19. Musango JK, Brent AC (2011) A conceptual framework for energy technology sustainability assessment. Energy Sustain Dev 15(1):84–91
20. Masui K, Sakao T, Kobayashi M, Inaba A (2003) Applying quality function deployment to environmentally conscious design. Int J Qual Reliab Manag 20(1):90–106
21. Mir MA, Ghazvinei PT, Sulaiman NMN, Basri NEA, Saheri S, Mahmood NZ, Jahan A, Begum RA, Aghamohammadi N (2016) Application of TOPSIS and VIKOR improved versions in a multi criteria decision analysis to develop an optimized municipal solid waste management model. J Environ Manag 166:109–115
22. Pope J, Annandale D, Morrison-Saunders A (2004) Conceptualising sustainability assessment. Environ Impact Assess Rev 24(6):595–616
23. Prato T (2007) Assessing ecosystem sustainability and management using fuzzy logic. Ecol Econ 61(1):171–177
24. Phillis YA, Andriantiatsaholiniaina LA (2001) Sustainability: an ill-defined concept and its assessment using fuzzy logic. Ecol Econ 37(3):435–456
25. Paju M, Heilala J, Hentula M, Heikkilä A, Johansson B, Leong S, Lyons K (2010) Framework and indicators for a sustainable manufacturing mapping methodology. In: Simulation conference (WSC), proceedings of the 2010 winter. IEEE, pp 3411–3422
26. Parkouhi SV, Ghadikolaei AS (2017) A resilience approach for supplier selection: using fuzzy analytic network process and grey VIKOR techniques. J Clean Prod 161:431–451
27. Ren J, Fedele A, Mason M, Manzardo A, Scipioni A (2013) Fuzzy multi-actor multi-criteria decision making for sustainability assessment of biomass-based technologies for hydrogen production. Int J Hydrog Energy 38(22):9111–9120
28. Romli A, Prickett P, Setchi R, Soe S (2015) Integrated eco-design decision-making for sustainable product development. Int J Prod Res 53(2):549–571
29. Székely F, Knirsch M (2005) Responsible leadership and corporate social responsibility: metrics for sustainable performance. Eur Manag J 23(6):628–647
30. Tseng ML (2013) Modeling sustainable production indicators with linguistic preferences. J Clean Prod 40:46–56
31. Tadić S, Zečević S, Krstić M (2014) A novel hybrid MCDM model based on fuzzy DEMATEL, fuzzy ANP and fuzzy VIKOR for city logistics concept selection. Expert Syst Appl 41(18):8112–8128

32. Vinodh S (2011) Assessment of sustainability using multi-grade fuzzy approach. Clean Technol Environ Policy 13(3):509–515
33. Vinodh S, Jayakrishna K (2011) Environmental impact minimisation in an automotive component using alternative materials and manufacturing processes. Mater Des 32(10):5082–5090
34. Waheed B, Khan F, Veitch B (2009) Linkage-based frameworks for sustainability assessment: making a case for driving force-pressure-state-exposure-effect-action (DPSEEA) frameworks. Sustainability 1(3):441–463
35. Wu YH, Ho CC (2015) Integration of green quality function deployment and fuzzy theory: a case study on green mobile phone design. J Clean Prod 108:271–280
36. Xu F, Liu J, Lin S, Yuan J (2017) A VIKOR-based approach for assessing the service performance of electric vehicle sharing programs: a case study in Beijing. J Clean Prod 148:254–267
37. Yang QZ, Chua BH, Song B (2009) A matrix evaluation model for sustainability assessment of manufacturing technologies. World Acad Sci Eng Technol 56:493–498
38. Zhao X, Huang K, Li G (2013) Manufacturing vendor selection based on cross-entropy measure with fuzzy VIKOR method. IFAC Proc Vol 46(9):1973–1978
39. Raut R, Kharat M, Kamble S, Kumar CS (2018) Sustainable evaluation and selection of potential third-party logistics (3PL) providers: an integrated MCDM approach. Benchmark: Int J 25(1):76–97
40. Sahu AK, Narang HK, Raiput MS, Sahu NK, Sahu AK (2018) Performance modeling and benchmarking of green supply chain management: an integrated fuzzy approach. Benchmark: Int J. https://doi.org/10.1108/BIJ-02-2017-0032
41. Sahu AK, Datta S, Mahapatra SS (2016) Evaluation and selection of resilient suppliers in fuzzy environment: exploration of fuzzy-VIKOR. Benchmark: Int J 23(3):651–673

Sustainability in Manufacturing

K. Raj Kumar Reddy, P. Kalpana, and A. Deiva Ganesh

Abstract The world population is constantly increasing, it is a concern as natural resources are limited and economic imbalances are going higher. Given the situation, it is required to promote sustainability to ensure resource conservation for coming generations. Manufacturing accounts for the largest energy consumption and making the manufacturing processes sustainable can lead to great results. Sustainability has three verticals, i.e., Environmental, Economical, and Social. Manufacturers should focus on all three verticals to announce themselves as sustainability adopters, in fact, most of them are trying to follow sustainability norms in their practices, but the practices lack structuredness. Recent surveys show that millennials are willing to pay extra for sustainable products. This acts as motivation for makers to focus on sustainability and promote it through their products and campaigns. The main challenge in this process is how to measure the sustainability index along with finding gaps in their activities. The next challenge is how to improve sustainability and maintain transparency in documenting those. Data is the new oil. It is trivial for any technology adoption. Industry 4.0 techniques exactly enable the same, it starts with acquiring data and ends with useful recommendations. In recent times, emerging technologies are being adopted across industries to make a positive difference in their operations. Technologies like internet of things, data science, artificial intelligence, blockchain technology, and cyber security can reduce uncertainties and improve efficiency. This chapter focuses on measuring sustainability and technology's role in ensuring sustainability in manufacturing, it throws light on Industry 4.0 automation techniques and how data can be leveraged to make a difference in manufacturing.

K. R. K. Reddy
Data Scientist, Chennai, Tamil Nadu, India

P. Kalpana (✉) · A. D. Ganesh
Department of Mechanical Engineering, Indian Institute of Information Technology Design & Manufacturing (IIITDM), Kancheepuram, Tamil Nadu, India
e-mail: kalpana@iiitdm.ac.in

A. D. Ganesh
e-mail: mdm20d002@iiitdm.ac.in

J. Kandasamy et al. (eds.), *Progress in Sustainable Manufacturing*, Management and Industrial Engineering, https://doi.org/10.1007/978-981-99-0201-9_5

Keywords Sustainability · Manufacturing · Technology · Industry 4.0 · Automotive manufacturing

1 Introduction

In the current modern era, things are changing rapidly, technologies are evolving within no time, and existing technologies won't be relevant if they are not adding up any value to the ever-evolving ecosystems. Also, the world is going through tough times like geopolitical tensions, pandemics transitioning into endemic, most nations are interdependent and related in a globalized world (exports and imports are part of any country's trade), and frequent occurrence of supply chain disruptions. On the other hand, the world is witnessing increased population density and economic inequality, which is driving supply and demand disparities and concerns over environmental conservation [1]. Here come sustainability compliances, which define how resources should be utilized and ecosystems must be built. At the end of the day it's up to responsible stakeholders to adopt and implement green and sustainable practices into the process for the greater good, as sustainable practices won't always lead to profit.

From the rock ages to the modern era, manufacturing remains constant as value addition and it will be trivial for the future too, as it is essential to fabricate and assemble the parts to make a product. So, making manufacturing processes a sustainable practice is very much essential to conserve the resources for the next generation and make manufacturing safer for the environment. Sustainable practices promote the usage of resources efficiently. They oppose exploitation in any manner including resources, environment, manpower, money, etc. According to reports, manufacturing processes utilize roughly 1/3rd of the world's energy, so small changes can promote sustainability in practices [2]. Manufacturing is of multiple varieties, starting from small to mammoth setups. At any instance manufacturing is resource-intensive. It requires high energy, manpower, materials, and money. Any change in granular level can result in big change, for example, if one can monitor tool wear and tear effectively, it will have a huge role in reducing energy consumption. In situations like these, technology comes into focus. With the use of advanced technologies one can monitor, understand, and act on their manufacturing practices, which eventually produces encouraging results [3]. In current times, data about processes and products are more essential for manufacturers and customers, as people say data is the new oil.

2 Sustainable Manufacturing

Sustainable manufacturing or green manufacturing exists to ensure the minimum waste, efficient use of resources, usage of natural resources, and preserving them for the generations to come [2]. Also, sustainable manufacturing focuses on promoting green initiatives, regulating the manufacturing practices, and ensuring minimum waste during production. To make sure of these responsibilities, emerging technologies' adoption is key. Nowadays more customers are interested in knowing the sustainability of the making process and the sustainability index of products and makers [4]. Post-pandemic, many are eager to encourage and engage in sustainable practices and products. Moreover, three out of four millennials are willing to pay extra for products and services if makers comment on improving the environment in a positive way [5]. So measuring sustainability is also a challenge and the same will be an opportunity if the measuring is done properly.

2.1 How to Measure the Sustainability

Sustainability must be measured to reflect on the sustainability index [4]. Sustainability needs to be measured from three perspectives, i.e., Environmental, Economical, and Social [6]. The detailed metrics are as follows:

1. **Environmental** aspects deal with the effectiveness of material usage, reduction of waste produced, energy consumption in the process, and effective usage of assets including tools and machines. If machine tools are not in desirable condition, it maximizes energy consumption, can increase waste, and cause damage as well, so it is essential to monitor machine tools in manufacturing processes in near real-time to ensure good conditions of machine tools and parts produced.
2. **Economical** aspects deal with the optimization of equipment and their cycle times, reduction of machine downtimes, and effective deployment of labor. Cost-saving should be done, but it shouldn't impact the environment and social sustainability of the firm.
3. **Social** aspects deal with the firm's contributions toward society, it can be corporate social responsibility (CSR) activities or ensuring no disturbances to society through manufacturing activities, i.e., Noise emission, disposing of waste without proper measures, etc. Social sustainability is mostly an overlooked aspect of sustainability among all [6], yet it measures the sustainability of an entity accurately.

Once sustainability is measured, the next important thing is how to improve the sustainability index and what are the practices to ensure things are in the right place.

2.2 Sustainable Manufacturing Practices

1. **Understanding and optimizing the utilization** of every grain of raw material, the units of power consumed, and the time spent by labor to finish a task is essential and valuable. So, planning the tasks and maximizing the utilization of assets and manpower is key to ensuring profitability and sustainability [5].
2. **Lowering pollution and reducing the amount of waste**: Ensuring the minimum waste and taking care of the waste produced, i.e., not letting the waste disturb the environment and cause any form of pollution [5].
3. **Recycling**: Whatever waste is produced should get disposed of efficiently and the waste particles should be recycled accordingly to ensure conservation of resources by efficiently using them. Also, manufacturers can print their product's raw material combination on top of their goods, so that they will get recycled after usage. For example, to dispose of batteries from electric vehicles, manufacturers can transparently print their raw material combination so that batteries get disposed of efficiently after usage [5].
4. **Adopt time saving techniques**: Firm management should be aware of the production lead time and machine tools' condition to perform the specified task. It can be done by understanding the condition of machines through data and the complexity of tasks they are going to perform [5].

Along with this management, integrity is trivial to ensure sustainability in manufacturing, they should adopt technology in their practices to ensure better and transparent results.

2.3 Management Integrity

Management integrity is required in adopting sustainable manufacturing, it is trivial to estimate manufacturing facilities' carbon profile to understand the baseline of environmental footprint. Companies should prepare their consumption data and enhance visibility from most of their manufacturing activities, to promote compliance in a transparent manner. Also, data should be prepared in a way that it can be audited, which can provide proof to clients and customers about their level of sustainability in all three verticals [6]. Management should brainstorm on resolutions, execute their thoughts by appointing corresponding people (New roles such as sustainability officer, addition to Quality, Production, Maintenance, and Safety people), and reflecting on the results to fine-tune the process even more.

3 Role of Technology

Adopting emerging technologies is necessary for ensuring efficient operations, and data is essential for any technology adoption. Industry 4.0 [7] encourages industries and people to automate their practices to generate data from their manufacturing processes, which can be used to understand the discrepancies in the process and establish visibility and transparency in manufacturing activities. This approach takes several steps to implement in a real-world manufacturing setup, which is cost-intensive, but on the same scale results can be obtained. Automation can be broadly categorized into three steps [3], i.e.,

1. **Machine Automation:** Automating the manufacturing equipment and establishing a way to acquire the data from machines. This is an equipment-rich step, which requires PLCs, computing devices, and acquisition and transmission devices to create the ecosystem.
2. **Information Automation:** Acquired data must be stored in centralized or decentralized databases, which also carry the analysis algorithms to examine the data and make decisions [14]. This step comprises decision automation and action automation as well.
3. **Knowledge Automation:** This step is far superior compared to others, with only a few requirements. It deals with incorporating knowledge into the machine, which provides autonomy of decision making to the machine itself. It involves rigorous analysis to build a robust process, otherwise, this can be destruction [8].

With machine automation in place, firms can acquire data through IoT devices and sensors from machines and manufacturing processes, then store the acquired data in databases. The data stored needs to be subjected to detailed analysis, only then the real value around data be unlocked. In most cases the acquired data is not subjected to detailed analysis, as it is being used for process log fulfillments and regular checks [3] (Fig. 1).

Coming to the technology stack, data science, machine learning, and artificial intelligence are at their peak in most of the applications [9]. Through these technologies predictions can be performed and intelligence can be incorporated into the machines. On the other hand, distributed ledger technologies such as Blockchain technology (BCT) are booming, which can couple trust, visibility, accountability, transparency, and traceability in manufacturing processes and finished goods [10]. Technology can be used to control and monitor manufacturing processes, and also to track input material consumption, machine condition, equipment wear and tear, energy usage, any discrepancies, etc. Adopting digital twin and digital thread frameworks can result in efficient overall equipment efficiency (OEE) [11]. The digital twin ensures the synchronization of physical setup with virtual representation; digital thread ensures the information delivery to the right place at the right time [11].

Manufacturing is the most happening industry, with machinery with advanced technologies being deployed to ensure the quality, accuracy, and precision of finished goods, which can be labeled as the primary indicator of customer satisfaction. With

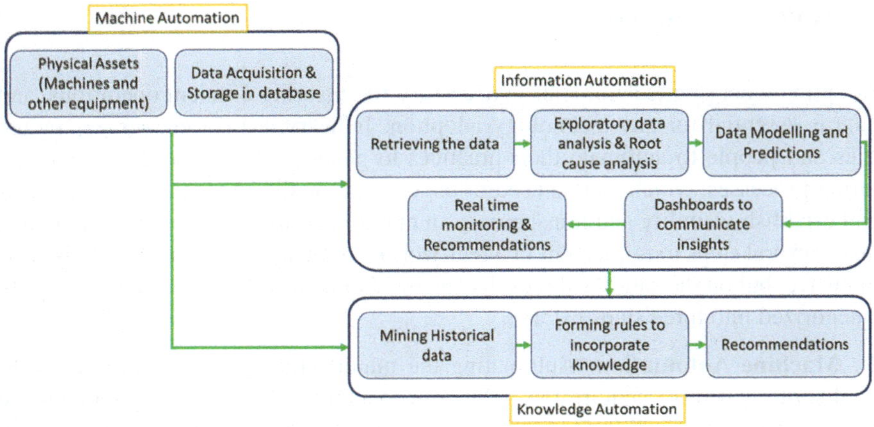

Fig. 1 Automation systems in manufacturing (industry 4.0)

the level of complexity in applications, sustainable practices can go deviated at times, but technology utilization for manufacturing practices can serve as a bridge. There is a debate around the sustainability of technologies themselves, many claim that crypto mining is not a sustainable practice, it is a challenge to draw a line, where technology should stop, and others take a front seat.

After technology, the next key thing in sustainable manufacturing is renewable energy [12]. As the name suggests renewable energy is energy collected from resources that are naturally replenished, it includes sunlight, wind, tide, waves, rains, thermal heat, etc. Promoting renewable resources in manufacturing can result in resource conservation, unlike fossil fuels, renewable sources are the most sustainable [12]. Through technology, one can maintain transparency in power purchasing agreements, and meet compliance to label the manufacturing setup as sustainable [10]. How much positive change renewable energy resources can bring is still a question mark.

3.1 Technology Use Cases in Manufacturing

Technology can be adopted with the right data from facilities available. Following are a few use cases of technology for sustainable manufacturing:

1. **Real-time monitoring of machinery and equipment:** Through data acquired, the firm must establish real-time monitoring systems to ensure the right condition and working of machines and smooth flow of manufacturing. The use of data visualization systems is trivial for employees to be informed about the progress
2. **Diagnostic analysis and root cause analysis**: The data must be utilized for drill-down analysis to understand and inspect every step of the manufacturing.

Through a detailed analysis, the firm can understand the critical problems and root causes causing those issues.

3. **Predictions on machine health, tool life, and quality and predictive and preventive maintenance**: With historical data, the firm should understand the risks machines can cause and maintain those in time to reduce their downtime. Through predictive and preventive maintenance firms can achieve unplanned machine stoppages. The difference between them is that predictive maintenance is done periodically, whereas preventive maintenance is scheduled at regular intervals to ensure the right conditions of machines [13]. Also, predictions must be performed on tool and machine life.
4. **Optimizing process route, cycle time:** Effective utilization needs optimization of sources and processes, so that efficiency goes up and waste produced goes down. Utilizing data, firms can focus on optimizing manufacturing tasks with existing infrastructure, and advanced optimization algorithms can facilitate efficient operations.
5. **Maintaining transparency on sustainability index**: Once the firms make efforts on sustainability toward environmental, economic, and social aspects, the activities and the impacts must be documented in a transparent manner, so that it acts as a front face to attract the customers. BCT can be leveraged to maintain transparency. BCT can be used to track sustainability activities efficiently and measure the sustainability index transparently [14].

Along with these other use cases also can be considered with the availability of data and criticality of the problem. Firms can utilize design thinking techniques to understand existing situations, root causes, and potentiality of ideas to solve problems.

4 Automotive Manufacturing

Let's dive into automotive manufacturing. It is one of the most complex manufacturing processes. It involves multiple suppliers who supply spare parts, transportation, and original equipment manufacturers (OEM) who make the finished product. Automotive manufacturing is resource, capital, and manpower intensive. A small difference in machine or process can lead to a lot of wastage of material and energy [10]. So, airtight manufacturing is required to ensure efficient usage of raw material and machines. The number of raw materials used in car manufacturing is mammoth. Recycling some of them to a certain extent can lead to very fruitful results and many can utilize those to fulfill their needs. OEMs are bound to follow circular economy [15] principles, i.e., reducing the waste being produced from manufacturing processes, and if waste is produced it must be recycled and reuse must be enabled. The steps in the circular economy are to eliminate or reduce waste produced from the process, enable circulation of used materials and products, and regenerate [15]. Circular economy enables sustainability in the manufacturing processes, as every raw material, component, and product carries a value in automotive manufacturing. OEMs must recover them at the end of their life.

On top of this, cars on the road do emit a lot of greenhouse gasses and a lot of carbon footprint. To address this issue manufacturers came up with environmentally friendly cars, i.e., Battery Electric vehicles (BEV), plug-in hybrid vehicles, and vehicles based on hydrogen fuel cells. Among these, BEVs are the most happening and attractive vehicles, due to the sustainable practices it is offering when it is on the road.

BEV manufacturing is no different from conventional vehicles, so the sustainability of BEV manufacturing is not a question, but the question is how sustainable the battery manufacturing is. The raw material for BEV can be cadmium, cobalt, lead, lithium, and nickel; these raw materials are not widely available and the demand for BEV is going up, causing mismatch in supply and demand. Most of these raw materials carry some side effects, so we can't term these as eco-friendly materials. BEVs may be built for environmental conservation and to reduce pollution levels, but usage of these materials won't assure the sustainability of the future.

5 Conclusion

Sustainability is very much relevant in ensuring the conservation of goods for coming generations. Since manufacturing consumes the world's 1/3rd amount of energy it is essential to promote sustainable manufacturing to conserve the resources. Moreover, post-pandemic there is huge interest from consumers toward sustainable products, also millennials are ready to pay extra for products which are promoting sustainability. To ensure sustainability in manufacturing, emerging technologies come as resourceful. Under Industry 4.0 initiatives, many industries set up machine automation systems, which acquire data from machines and manufacturing processes. But that data is rarely utilized fully to make the real difference, it was being used mostly for occasional checks. Through the data, most of the emerging technologies like data science, AI, Machine learning, and BCT can be leveraged to ensure the right conditions of machines, reduction in waste, reduction in lead time, etc. Automation is of three levels, i.e., machine automation, information automation, and knowledge automation. Based on industry and scope the level of automation can be achieved. In this chapter, we suggested the verticals to measure sustainability and ways to improve sustainability through emerging technologies. Here, only conceptual discussions are being presented, with the existing infrastructure, firms should look at things from a sustainability perspective, so the next steps for this context are leveraging the discussion here to practical implementation.

References

1. Tiwari S, Wee HM, Daryanto Y (2018) Big data analytics in supply chain management between 2010 and 2016: insights to industries. Comput Ind Eng 115:319–330
2. Sustainable manufacturing. https://www.epa.gov/sustainability
3. Ge Z, Song Z, Ding SX, Huang B Data Mining and Analytics in the Process Industry: The Role of Machine Learning
4. Lucato WC, Santos JC, Pacchini AP. Measuring the sustainability of a manufacturing process: a conceptual framework
5. Ways to achieve more sustainable manufacturing practices. https://www.thepowerscompany.com/maintenance-performance/5-ways-to-achieve-more-sustainable-manufacturing-practices/
6. Social sustainability. https://www.adecesg.com/resources/faq/what-is-social-sustainability/
7. Reddy KR, Kalpana P (2021) Chapter 11 impact of COVID-19 on global supply chains and the role of digitalization: A VUCA approach. Springer Science and Business Media LLC
8. Zhiqiang W, Jun L (2017) A review of object detection based on convolutional neural network. IEEE, pp 11104–11109
9. Attaran M, Gunasekaran A (2019) Blockchain-enabled technology: the emerging technology set to reshape and decentralise many industries. Int J Appl Decis Sci 12(4):424–444
10. Reddy KR, Gunasekaran A, Kalpana P, Sreedharan VR, Kumar SA (2021) Developing a blockchain framework for the automotive supply chain: a systematic review. Comput Ind Eng 157:107334. ISSN: 0360-8352. https://doi.org/10.1016/j.cie.2021.107334
11. Digital twin and Digital Thread. https://www.challenge.org/insights/digital-twin-and-digital-thread/
12. Renewable energy. https://www.energy.gov/eere/renewable-energy
13. Predictive and preventive maintenance. https://www.artesis.com/predictive-vs-preventive-maintenance/
14. Alladi T, Chamola V, Parizi RM, Choo KR (2019) Blockchain applications for industry 4.0 and industrial IoT: a review. IEEE Access 7:176935–176951
15. Circular economy. https://ellenmacarthurfoundation.org/topics/circular-economy-introduction/overview

Experimental Investigation of Machining NIMONIC 80 Alloy by WEDM Process via Multi-objective Optimisation Techniques: A Sustainable Approach

Bikash Ranjan Moharana, Bikash Chandra Behera, Shoeb Ahmed Syed, Kamalakanta Muduli, and Shubham Barnwal

Abstract Sustainable machining should address manufacturing sustainability. Sustainable means something can be sustained, or continue at the same level, in principle forever. Sustainable manufacturing is the production of environmentally friendly, energy-and resource-efficient, and safely produced items. This term incorporates sustainable and eco-friendly product manufacturing. Sustainability in machining is the creation of items (components, etc.) using a non-polluting, minimal, and conserving subtractive cutting method. Conventional machining techniques are being phased out in favour of non-traditional machining processes due to increased demand for high surface polish and complicated form geometries. Wire EDM is a non-traditional machining method. Surface roughness, MRR, and kerf width are crucial in machining. This project summarises the Taguchi optimisation strategy, utility method, and confirmation test to optimise Wire EDM cutting parameters for NIMONIC 80 ALLOY. Optimisation aims to minimise Kerf breadth while maximising MRR and surface quality. NIMONIC 80 ALLOY is used in this work. Die electric fluid is DI water, and the tool is 0.25 mm brass wire. This project uses Taguchi's L9 orthogonal array. Optimising pulse on, pulse off, wire feed, and feed wire tension. MRR, surface roughness, and Kerf were assessed. Multiple outputs make it difficult to choose a single ideal parameter setting. Taguchi-PSI and Gray Relational Analysis as the multi-optimisation approach are used to reduce ambiguity.

Keywords NIMONIC 80 Alloy · WEDM · Taguchi-PSI · GRA

B. R. Moharana · B. C. Behera · S. Barnwal
Mechanical Engineering Department, CV Raman Global University, Bhubaneswar, India
e-mail: bikash@niser.ac.in

S. A. Syed · K. Muduli (✉)
Mechanical Engineering Department, Papua New Guinea University of Technology, Lae, Papua New Guinea
e-mail: kamalakantam@gmail.com

S. A. Syed
e-mail: shoeb.syed@pnguot.ac.pg

J. Kandasamy et al. (eds.), *Progress in Sustainable Manufacturing*, Management and Industrial Engineering, https://doi.org/10.1007/978-981-99-0201-9_6

1 Introduction

Sustainability in manufacturing should be considered while defining sustainable machining. Sustainable suggests that something can be maintained, i.e., continuing at the same level, in theory forever, according to the article on "sustainability." Production of manufactured goods that are environmentally friendly [1], save energy and natural resources, and are safe for the people who work with them and those who buy their products is considered "sustainable manufacturing." This definition includes the production of sustainable products as well as the production of any product in an environmentally friendly manner. Sustainability in machining can be defined as the production of products (components, etc.) through a subtractive process based on cutting (material removal through the cutting action with a machine tool to create surfaces and features) in a way that is non-polluting, minimises and conserves; this definition applies to machining [1]. Conventional machining processes are currently being phased out in favour of non-traditional machining processes as a direct result of an increase in the demand for a high surface finish as well as the machining of complex shape geometries. One of the processes that is considered to be non-traditional form of machining is known as wire electrical discharge machining (WEDM) [2].

The most effective and efficient non-traditional method of material removal is known as WEDM. This method has been put to extensive use in the current metal working industries for the production of complicated profiles in dies and moulds used in tool and die manufacture as well as other sectors such as aerospace, automotive, surgical component manufacturing, and so on. However, there are troublesome aspects that need to be properly explored so that the functioning precision may be improved. With wire electrical discharge machining (WEDM), the economics of machining and the pace of production are determined by the material removal rate. When configuring the machining parameters, the primary objective is to achieve the highest possible MRR and SF while maintaining the narrowest possible kerf width. The process variables notably electrical variables are more critical elements in choosing response values. It is required to forecast the response values of the machining process in order to select which input parameters should be provided in order to achieve precise and accurate machining.

Because this is an expensive operation, the optimal selection of process parameters is very vital in order to significantly enhance output rate while simultaneously decreasing the amount of time spent on machining. Material removal rate (MRR), kerf width, and surface roughness are the three most essential output metrics that determine how well machining is done. In natural settings, the rate at which material was removed from surfaces rose as the roughness of those surfaces decreased, and vice versa.

When compared to the traditional electrical discharge machining (EDM) technique, which relies on an electrode to kick off the sparking procedure, the wire electrical discharge machining (WEDM) method is regarded as an innovative modification. On the other hand, wire electrode discharge machining (WEDM) makes use of a wire electrode that moves constantly and is composed of thin copper, brass, or tungsten and is able to achieve extremely tiny corner radii. A mechanical tensioning mechanism is used to keep the wire in tension, which reduces the likelihood of generating components that are erroneous. Because there is no direct contact between the work piece and the wire during the WEDM process, the mechanical stresses that would normally occur during machining are completely eliminated. The material is eroded in front of the wire.

Since it was initially put to use more than three decades ago, the technique known as wire electrical discharge machining (WEDM) has seen enormous development. In 1974, D. H. Dulebohn created an optical line follower system for the purpose of autonomously controlling the form of components that were going to be machined using a WEDM process. By 1975, the industry had a greater understanding of the method as well as the possibilities it offered, which led to a significant expansion in the practice's popularity. It wasn't until the late 1970s that the computer numerical control (CNC) technology was first implemented into WEDM. Because of its extensive capabilities, it has been able to expand into the manufacturing industry, as well as the aerospace and automotive sectors, and almost all fields of machining conductive materials. This is due to the fact that WEDM offers the greatest choice, and in some cases the only alternative, for machining conductive, exotic, high strength, and temperature resistant materials, as well as conductive engineered ceramics, with the capability of producing complicated forms and profiles.

WEDM has a huge amount of promise in terms of its use in the modern day metal cutting business. This is because it can achieve significant dimensional accuracy, surface polish, and contour generating aspects of goods or components. In addition, the cost of the wire only accounts for ten per cent of the total operational costs of the WEDM process. WEDM allows for the avoidance of the challenges that are typical of die sinking EDM by substituting the complicated design tool with a moving conductive wire and the relative movement of a wire guide. This allows for the elimination of the issues.

The wire electrical discharge machining (WEDM) machine tool consists of a main worktable (X–Y), an auxiliary table (U-V), and a wire driving mechanism. The work piece is clamped on the main worktable. The servo motors for the D.C. drive system are responsible for moving the main table along the X and Y axes. The travelling wire is continually supplied from the wire feed spool and collected on the pick up spool as it goes through the work piece. The travelling wire is maintained under tension between a pair of wire guides that are situated on opposing sides of the work piece. The lower wire guide is fixed in place, whereas the upper wire guide, which is supported by the U-V table, may be moved transversely along the U and V axes in relation to the lower wire guide. By rotating the quill, it is also possible to set the top wire guide in a vertical orientation along the Z axis.

In order to electro erode the material of the work piece, a succession of electrical pulses that are created by the pulse generator unit are applied between the work piece and the travelling wire electrode. The X–Y controller moves the worktable, which is carrying the work piece, in a transverse direction along a route that has been defined and programmed into the controller as the process continues. The machining zone is continually cleansed with water that is going through the nozzle on both sides of the work piece as the machining process is carried out in a continuous fashion. It is essential that water does not ionise because of its role as a dielectric medium; this requirement must be strictly adhered to. Therefore, in order to maintain the conductivity of water and avoid the ionisation of water, an ion exchange resin is employed in the dielectric distribution system. This resin is utilised to maintain the conductivity of water.

The wire electrode has to be angled in order to generate taper machining. This can only be done by hand. This is accomplished by moving the top wire guide in relation to the lower wire guide along the U-V axis. Through simultaneous control of the movement of the X–Y table and the U-V table along their respective specified pathways that are recorded in the controller, the required taper angle may be attained. The NC programme provides the controller with the route information of the X–Y table as well as the U-V table in the form of linear and circular components. Displays a diagrammatic representation of the fundamental concept behind the WEDM method.

In wire electrical discharge machining, the mechanism of metal removal primarily involves the removal of material due to melting and vaporisation caused by the electric spark discharge generated by a pulsating direct current power supply between the electrodes. This discharge causes the electric spark. In WEDM, the negative electrode is represented by a wire that is constantly moving, while the positive electrode is represented by a dielectric liquid. WEDM makes use of water as a dielectric because of the low viscosity of water and the quick cooling rate of water.

It has not been determined if a particular theory can adequately explain the intricate machining procedure. Empirical data, on the other hand, reveals that the applied voltage forms an ionised channel between the closest points of the work piece and the wire electrodes in the early stage. This occurs in the initial stage. In the subsequent step, the real discharge happens, and it involves a substantial flow of current. At the same time, the resistance of the ionised channel steadily reduces. The high strength of the current continues to progressively ionise the channel, which results in the generation of a large magnetic field. Because of the compression caused by this magnetic field in the ionised channel, localised heating is produced. Even with sparks that only last for a very little period of time, the temperature of the electrodes may locally increase to extremely high values, which can be higher than the material's melting point. This happens because the kinetic energy of the electrons is converted into heat throughout the process. As a result of the high energy density, a portion of the material on both the wire and the work piece is eroded away by the process of locally melting and vaporising, and this is the predominate form of thermal erosion.

2 Literature Review

In preparation for this body of work, we have conducted a comprehensive literature review. In their study, Karsh and Singh [3] and colleagues found that the pulse on time and the pulse off time both had a substantial impact on the cutting rate. According to the results of the analysis of variance performed on the raw data, regulating the cutting rate requires a contribution of 68.15% from the pulse on time and 20.57% from the pulse off time. The cutting rate is raised either when the pulse on time is increased or when the pulse off time is decreased. The peak current doesn't have much of an impact on the cutting rate. Kumar and Chauhan [4] conducted an experimental examination to evaluate the influence of several factors on the surface roughness of NIMONIC 90. These parameters were peak current, pulse on time, pulse off time, servo voltage, and wire feed. Based on the findings of the experiment conducted while utilising by changing only one variable at a time The surface roughness is affected more noticeably by the peak current, pulse on time, and pulse off time. The influence of WEDM parameters such as pulse on time, pulse off time, peak current, servo voltage, and wire feed on surface roughness was studied by [5]. In order to conduct an analysis of the outcome for SR, they rely on statistical methods. A portion of the study conducted on several different materials. The WEDM machining of high carbon high chromium was carried out by Bhatia et al. in 2017 utilising brass wire. There is an application of the Taguchi single answer optimisation approach. The Surface Roughness is calculated for each of the input parameters, such as the peak current, the on time, the off time, and the wire tension. The difference in the values of SR that were anticipated and those that were actually measured is 3.93%. The pulse off time has the most significant impact on SR. Input parameters pulse on time and pulse off time are both set to 131 μs, and peak current is set to 220 A in order to produce the minimal SR using the Taguchi approach.

3 Experimentation Methodology

3.1 Selection of Machining Conditions

In the present study, the NIMONIC 80A super alloy was selected for the machining. Basically, the main compositions of NIMONIC alloys are nickel (Ni) and chromium (Cr). These alloys are well-known for their high-temperature, low-creep, and high performance. The dimension of the work piece material used in this study, i.e., NIMONIC 80A was 150 mm $\times$ 150 mm $\times$ 11 mm. Brass wire ($CuZn_{37}$) of diameter 250 μm was used as the electrode material. The brass wire is an alloy of zinc (37%) & copper (63%), having significant properties like relatively low cost compared to other electrode wires like copper, good mechanical properties, low conductivity, improved flush-ability, and higher tensile strength. The chemical compositions of work piece material and electrode wire are given in Tables 1 and 2, respectively. All the present experiments were conducted by ECOCUT® CNC wire-cut EDM machine using distilled water as dielectric fluid as shown in Fig. 1. The wire electrode, i.e., brass ($CuZn_{37}$) used during machining is shown in Fig. 2.

From the extensive literature survey, it has been found that the several machining inputs affects the machining performance during WEDM process [6–8]. So, a controlled factor window should be carefully chosen for the improvement of the

Table 1 Chemical Composition of NIMONIC 80a

Element	Ni	Cr	Co	Mo	Ti	Fe	C	Others
Weight %	51.43	20	19.5	5.6	2.2	0.25	0.043	0.997

Table 2 Chemical composition of tool wire ($CuZn_{37}$)

Cu (%)	Zn (%)	Sn (%)	P (%)	Fe (%)	Al (%)	Ni (%)
63	36.5	0.12	0.02	0.01	0.05	0.27

Fig. 1 ECOCUT® CNC wire-EDM machine set-up

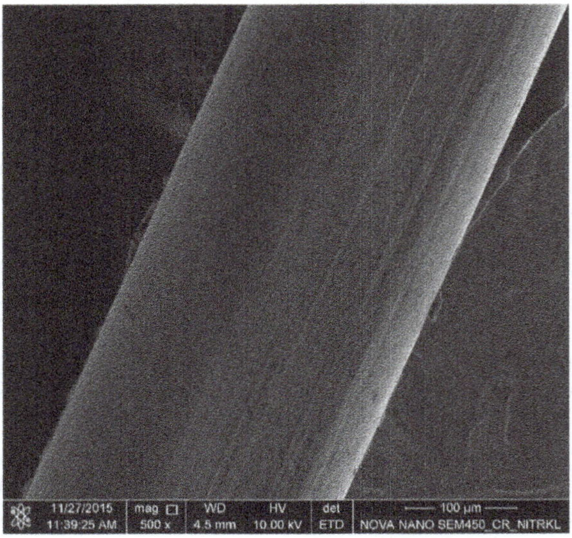

Fig. 2 Brass electrode wire ($CuZn_{37}$) used during machining

Table 3 Process parameters with experimental design levels

Parameters	UNITS	Symbol	Levels		
			L1	L2	L3
Pulse on time	μs	T_{on}	110	115	120
Pulse off time	μs	T_{off}	25	30	35
Wire feed	m/min	WF	10	12	14
Wire tension	N	WT	8	9	10

machining performance. On the basis of literature survey and WEDM machine set-up constraints, the input machining parameters with their range are shown in Table 3. The input machine parameters included in the experimentation, i.e., Pulse on time (T_{on}), Pulse off time (T_{off}), wire feed rate (WF), and wire tension (WT), whereas the material removal rate (MRR), surface roughness (Ra), and kerf width (Kf) are considered as the output responses, i.e., WEDM performance measures.

3.2 *Experimental Design*

The experiments are conducted based on Taguchi orthogonal array design matrix. The advantage of Taguchi method over the other methods is that the number and levels of technological parameters involved in the experimental design matrix is large whereas the number of experiments is minimal which leads to reduction in experimental costs and the time of experiment. In this present research, L9 orthogonal design matrix (shown in Table 4) is selected for experimentation by taking 3 input parameters with 3 different levels of variations as given in Table 3.

Table 4 Experimental design matrix with output responses

Expt. no	Pulse on time (μs)	Pulse off time (μs)	Wire feed (m/min)	Wire tension (N)	MRR (mm^3/min)	Ra (μm)	Kf (mm)
1	110	25	10	8	3.8949	2.052	0.362
2	110	30	12	9	2.8205	1.937	0.310
3	110	35	14	10	3.2478	1.819	0.373
4	115	25	12	10	4.8835	2.824	0.295
5	115	30	14	8	3.7728	2.614	0.287
6	115	35	10	9	3.6874	2.605	0.284
7	120	25	14	9	7.2405	3.270	0.365
8	120	30	10	10	5.9096	3.120	0.332
9	120	35	12	8	5.2503	3.111	0.315

The experimentation is carried out by following standard procedure as manually setting the appropriate pulse on time, pulse off time, wire feed and wire tension. A square slot having dimension 10 mm × 10 mm × 11 mm was shaped on the work piece, the corresponding machining time was noted and final weight of the work piece was taken. The same procedure was followed for the whole set of experiments. After completion of all experimentation, three output responses are considered as the evaluator of the machining characteristics such as MRR, Ra, and Kf. As per the machining characteristics, MRR should be maximum whereas Ra and Kf should be minimum. Subsequently, these scenarios fall under multi-objective optimisation.

- MRR is defined as the volume of the material removal per unit time and formulated below (1):

$$\text{MRR} = \frac{W_b - W_a}{\rho \times t}(\text{mm}^3/\text{min}) \tag{1}$$

where,

W_b and W_a	weights of work piece in grams before and after machining, respectively
ρ	density of work piece material (0.00819 g/mm^3)
t	machining time in minutes

- The kerf width is calculated as formulated below (2):

$$\text{KF} = \frac{\text{MRR}}{\text{CR} \times T}\ (\text{mm}) \tag{2}$$

where

CR	cutting rate in mm/min
T	thickness (11 mm) of work piece material.

- The average surface roughness (SR) was measured by using the Mitutoyo surface tester.

The designing of mathematical and computational equations to assist implementers is based on multi-objective prospects to make appropriate decisions on the importance of each indicator selected in the study. The multi-objective selection evaluation considers the selection, prioritisation, precedence order, and choice of appropriate solutions [9, 10]. The determination of precedence degrees in the parameters of the survey parameter is very necessary in making decisions.

3.3 Methodology for Multi Objective Optimisation by Taguchi-PSI

The PSI method used to determine different criteria is used in the following steps:

Step 1 (Problem definition): Determine the appropriate goals and be aware of the important aspects as well as potential solutions that are associated with the current decision-making circumstance.

Step 2 (Decision matrix formulation): Construct the first iteration of the decision matrix using the criteria that were chosen. If there are n total targets and m total experiments, then the decision matrix may be written by the equation "*m n*," where m and n are the respective numbers of experiments and targets (3).

$$X = \begin{bmatrix} x_{11} & x_{12} & \cdot & x_{1j} & x_{1n} \\ x_{21} & x_{12} & \cdot & x_{1j} & x_{1n} \\ \cdot & \cdot & \cdot & \cdot & \cdot \\ x_{i1} & x_{i2} & \cdot & x_{ij} & x_{in} \\ \cdot & \cdot & \cdot & \cdot & \cdot \\ x_{m1} & x_{m2} & \cdot & x_{mj} & x_{mn} \end{bmatrix} \tag{3}$$

Step 3 (Data Normalisation): In solving multi-objective problems the indicators need to be dimensionless quantity. For this purpose, indicators' values need to be normalised varying from 0 to 1 [11]. In this step, the experimental results were first normalised as per the consideration of the response characteristics. "Larger values" are desirable for advantageous type qualities whereas "Lesser values" for non-beneficial type qualities, which may be normalised by following Eqs. 4 and 5, respectively.

$$N_{ij} = \frac{x_{ij}}{x_{ij}^{max}} \tag{4}$$

$$N_{ij} = \frac{x_{ij}^{min}}{x_{ij}} \tag{5}$$

where, x_{ij} is the value of the indicators in the *i*th row and *j*th column ($i = 1, 2, 3, \ldots.., m$ and $j = 1, 2, 3, \ldots., n$). The decision matrix is standardised by Eqs. (4) and (5), it depends on the findings of the problem.

Step 4 (Mean value of the normalised data): For each characteristic the mean value of the normalised data may be determined as follows (6):

$$\overline{N} = \frac{1}{n}\sum_{i=1}^{n} N_{ij} \tag{6}$$

Step 5 (Compute the preference variation value): The preference variation value among the values of each indicator is calculated using Eq. (7):

$$\phi_j = \sum_{i=1}^{n} \left(N_{ij} - \bar{N}\right)^2 \tag{7}$$

Step 6 (Determine the deviation in preference value): The deviation of the priority level relative to each criterion is computed using Eq. (8):

$$\sigma_j = 1 - \phi_j \tag{8}$$

Step 7 (Compute the overall preference value): The overall priority value is determined for each attribute by using Eq. (9):

$$W_j = \frac{\sigma_j}{\sum_{j=1}^{m} \sigma_j} \tag{9}$$

Furthermore, it must be ensured that the value of the total overall preference must satisfy Eq. (10).

$$\sum_{j=1}^{m} W_j = 1 \tag{10}$$

Step 8 (Compute the preference selection index): The preferred selection index (PSI) is calculated for each alternative by using Eq. (11).

$$\theta_j = \sum_{j=1}^{m} N_{ij} \times W_j \tag{11}$$

Step 9 (Selection of appropriate alternatives for given application): Finally, the ranks are assigned to each choice according to the reducing value of the precedence index (θ_j). The highest preferred selection index value will be ranked first, and it will be the optimal solution.

3.4 Methodology for Multi-objective Optimisation by Grey Relational Analysis (GRA)

The GRA has been used for multi-response optimisation. It gives an effective solution to the ambiguity, multi-input, and isolated data problem. The relationship between input parameters and outputs can be established using GRA. The various steps followed in the GRA are shown.

In this methodology, the experimental outputs are first normalised based upon the response characteristics as follows:

For "higher the better" characteristics:

$$x_i(k) = \frac{\max y_i(k) - y_i(k)}{\max y_i(k) - \min y_i(k)} \tag{12}$$

For "lower the better" characteristics:

$$x_i(k) = \frac{y_i(k) - \min y_i(k)}{\max y_i(k) - \min y_i(k)} \tag{13}$$

where,

$x_i(k)$ and $y_i(k)$ normalised and experimental values for ith experiment using kth response, respectively,

$\min y_i(k)$ and $\max y_i(k)$ are the minimum and maximum values of $y_i(k)$ for the kth response, respectively.

Further, the Grey Relation Coefficient (GRC) "$\varphi_i(k)$" for the kth response characteristics in the ith experiment was calculated as follows (14).

$$\varphi_i(k) = \frac{\Delta\min + \varphi\Delta\max}{\Delta i(k) + \varphi\Delta\max} \tag{14}$$

where,

"$\Delta i(k)$" is the deviation sequence of the reference sequence and "φ" is the identification coefficient that lies in between 0 and 1.

"$\Delta\max$" and "$\Delta\min$" are the global maximum and minimum values of each arrangement correspondingly.

Then in the final step, the Grey Relational Grade (GRG) "γ" can be calculated using Eq. (15).

$$\gamma = \frac{1}{n}\sum_{i=1}^{n}\varphi_i(k) \tag{15}$$

4 Results and Discussion

All the experimentations were conducted as per the Taguchi orthogonal design matrix as described earlier. After successful machining operation, the respective selected output responses were calculated accordingly as given in Table 3. In search of optimal condition for machining, two different multi-optimisation techniques such as Taguchi-PSI and GRA are applied in this analysis.

4.1 Taguchi-PSI Approach

First of all, the decision matrix was formulated based upon the selected criteria, followed by normalisation to generate the normalised matrix considering Eqs. 4 and 5. Equal weightage was assigned to each criterion. Further, the mean value of normalised data, preference variation value, deviation in preference value and overall preference value were determined referring Eqs. 6, 7, 8, and 9, respectively. Finally, the preference selection index (PSI) was computed by following Eq. 11. The PSI values along with their respective ranking is presented in Table 5.

The response table and optimal parametric setting drawn from Taguchi-PSI analysis is shown in Table 6 and Fig. 3, respectively. By considering the "Larger the better" criterion, the optimal set of process parameters is "115-25-12-9". The probability plot of PSI (shown in Fig. 4) describes the fitness of the model. Using such findings, it was recommended that the pulse on time has the greatest contribution followed by wire tension, pulse off time, and wire feed (T_{on} > WT > T_{off} > WF) (Table 7).

Table 5 Preference selection index (PSI) values with ranking

Expt. no	Pulse on time (μs)	Pulse off time (μs)	Wire feed (m/min)	Wire tension (N)	PSI	Rank
1	110	25	10	8	0.7464	9
2	110	30	12	9	0.7717	2
3	110	35	14	10	0.7484	7
4	115	25	12	10	0.7765	1
5	115	30	14	8	0.7594	5
6	115	35	10	9	0.7608	4
7	120	25	14	9	0.7703	3
8	120	30	10	10	0.7560	6
9	120	35	12	8	0.7482	8

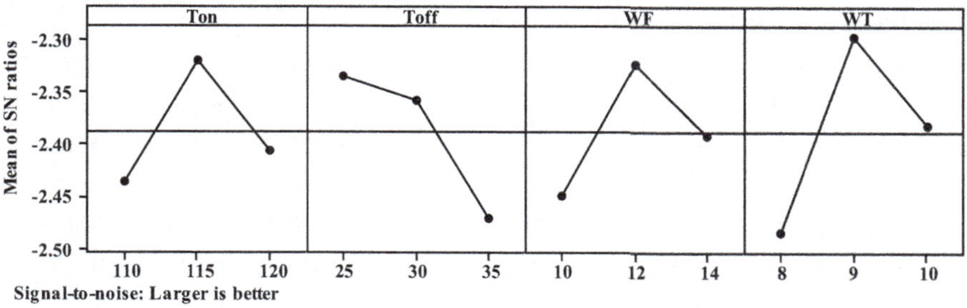

Fig. 3 Main effect plot for S/N ratios (PSI)

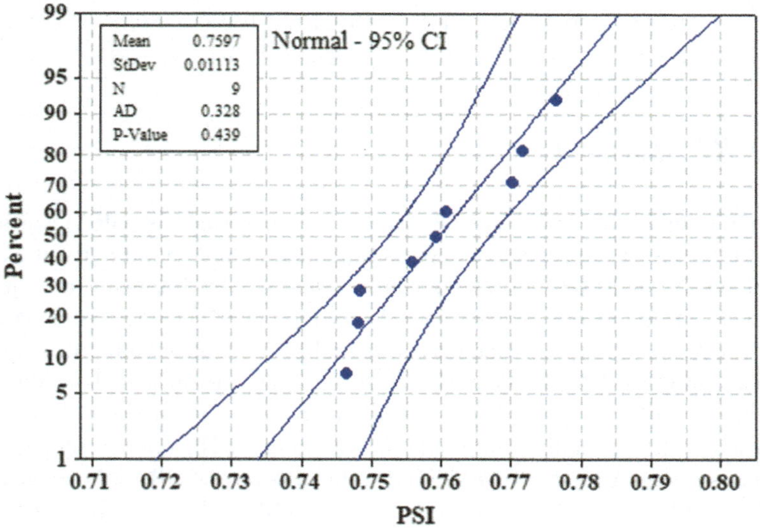

Fig. 4 Probability plot of PSI

Table 6 Response table for S/N ration (PSI)

Level	T_{on}	T_{off}	WF	WT
1	−2.436	−2.335	−2.448	−2.484
2	−2.321	−2.357	−2.323	−2.297
3	−2.406	−2.471	−2.392	−2.382
Delta	0.166	0.136	0.125	0.146
Rank	1	3	4	2

Table 7 GRC and GRG calculation from experimental results

Expt. no	Normalised results			Grey relational coefficient (GRC)			GRG
	MRR	Ra	Kf	MRR	Ra	Kf	
1	0.243	0.839	0.124	0.398	0.757	0.363	0.506
2	0.000	0.919	0.708	0.333	0.860	0.631	0.608
3	0.097	1.000	0.000	0.356	1.000	0.333	0.563
4	0.469	0.307	0.876	0.485	0.419	0.802	0.569
5	0.215	0.452	0.966	0.389	0.477	0.937	0.601
6	0.196	0.458	1.000	0.383	0.480	1.000	0.621
7	1.000	0.000	0.090	1.000	0.333	0.355	0.563
8	0.699	0.103	0.461	0.624	0.358	0.481	0.488
9	0.550	0.110	0.652	0.526	0.360	0.589	0.492

4.2 GRA Approach

The multi optimisation technique, i.e., Grey Relational Analysis (GRA) was implemented by following appropriate steps as explained in Sect. 3.4. Firstly, the experimental results of MRR, Ra, and Kf are normalised by following the "higher-the-better" characteristics (MRR) and "lower-the-better" characteristics (Ra and Kf). Then these normalised data are considered for GRC calculation of respective responses. Afterward, the GRG is estimated from GRC and presented in Table 8. By following the GRG rule, "Higher is better" policy was computed for all the correspondence [12]. The main effect plot is shown in Fig. 5. The optimal set of process parameters is getting "115-30-14-9". The probability plot of GRA (shown in Fig. 6) describes the fitness of the model. Like Taguchi-PSI model, the GRA model also recommended that the pulse on time has the greatest contribution followed by wire tension, wire feed, and pulse off time (T_{on} > WT > WF > T_{off}).

Table 8 Response table for S/N ration (GRA)

Level	T_{on}	T_{off}	WF	WT
1	−5.076	−5.268	−5.429	−5.500
2	−4.486	−4.992	−5.127	−4.483
3	−5.794	−5.096	−4.801	−5.373
Delta	1.308	0.276	0.628	1.017
Rank	1	4	3	2

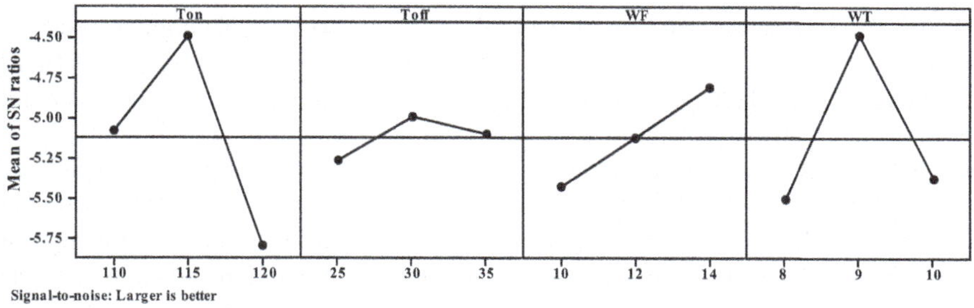

Fig. 5 Main effect plot for S/N ratios (GRA)

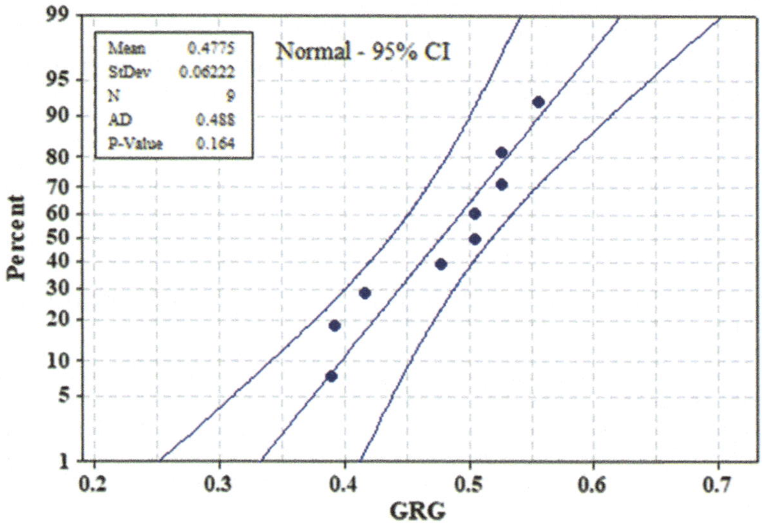

Fig. 6 Probability plot of GRA

5 Conclusions

In the present work, the impact of machine settings on MRR, Ra, and Kf during machining of NIMONIC 80 alloy using WEDM method has been studied. By adopting L9 Taguchi orthogonal array, the tests are carried out. Further, the experimental findings were examined by adopting several multi criteria decision making (MCDM) methodologies to determine the ideal parametric settings to prevent the provocative aims like larger MRR, lower Ra, and lower Kf. The following findings are stated below as follows:

- We have been successful in implementing a more sustainable method for WEDM of NIMONIC 80 alloy.

- The Taguchi orthogonal array-based L9 design matrix was taken into consideration for this work in order to investigate the impact of MRR, Ra, and Kf of WEDM machining on NIMONIC 80 alloy using brass wire electrode and de-ionised water as the di-electric environment.
- The MCDM technique, also known as the Taguchi-PSI findings, indicates that a high MRR, a low Ra, and a low Kf may be achieved through WEDM processing of NIMONIC 80 alloy similar to the 4th experimental run, which had the highest grade of 0.7765. The sequential best parametric combination, which was determined to be T_{on}: 115-T_{off}:25-WF:12-WT:10, was discovered, and the findings also showed that "T_{on}" seemed to be the most important factor, followed by WT, T_{off}, and WT.
- In a similar manner, the GRA method likewise confirms that the "T_{on}" is the most significant component when compared to all of the others.
- The probability plot provides further evidence that the design model is adequate for any of the two MCDM methodologies.

References

1. Chetan SG, Venkateswara Rao P (2015) Application of sustainable techniques in metal cutting for enhanced machinability: a review. J Clean Prod 100:17–34
2. Jawahir IS, Schoop J, Kaynak Y, Balaji AK, Ghosh R, Lu T (2020) Progress toward modeling and optimization of sustainable machining processes. J Manuf Sci Eng 142(11)
3. Karsh PK, Singh H (2018) Multi-characteristic optimization in wire electrical discharge machining of inconel-625 by using taguchi-grey relational analysis (GRA) approach: optimization of an existing component/product for better quality at a lower cost. In: Products KK, Davim JP (eds) Design and optimization of mechanical engineering Hershey. IGI Global, PA, USA, pp 281–303
4. Kumar R, Chauhan S (2015) Study on surface roughness measurement for turning of Al 7075/10/SiCp and Al 7075 hybrid composites by using response surface methodology (RSM) and artificial neural networking (ANN). Measurement 65:166–180
5. Pal RK, Garg H, Sarepaka RV, Karar V (2016) Experimental investigation of material removal and surface roughness during optical glass polishing. Mater Manuf Processes 31(12):1613–1620
6. Manjaiah M, Narendranath S, Basavarajappa S (2016) Wire electro discharge machining performance of TiNiCu shape memory alloy. Silicon 8(3):467–475
7. Soepangkat BOP, Pramujati B (2013) Optimization of surface roughness and recast layer thickness in the wire-EDM process of AISI D2 tool steel using Taguchi-grey-fuzzy. In: Applied mechanics and materials, vol 393. Trans Tech Publications Ltd., pp 21–28
8. Singh T, Misra JP, Singh B (2017) Experimental investigation of influence of process parameters on MRR during WEDM of Al6063 alloy. Mater Today: Proc 4(2):2242–2247
9. Biswal JN, Muduli K, Satapathy S (2017) Critical analysis of drivers and barriers of sustainable supply chain management in Indian thermal sector. Int J Procure Manag 10(4):411–430
10. Muduli K, Barve A (2013) Modelling the behavioural factors of green supply chain management implementation in mining industries in Indian scenario. Asian J Manag Sci Appl 1(1):26–49
11. Biswal JN, Muduli K, Satapathy S, Yadav DK (2019) A TISM based study of SSCM enablers: an Indian coal-fired thermal power plant perspective. Int J Syst Assur Eng Manag 10(1):126–141
12. Tosun N, Pihtili H (2010) Gray relational analysis of performance characteristics in MQL milling of 7075 Al alloy. Int J Adv Manuf Technol 46(5):509–515

Prediction and Optimization of Sustainable Production Processes for Automotive Components

Avinaash Jaganaa, Voonna Balakrishna Taruna, Koduri Naga Ganapathi Lakshmi Reshawantha, G. Rajyalakshmi, and K. Jayakrishna

Abstract This paper includes the study of power train components of an All-Terrain Vehicle. The research aims to optimize the designs and reduce the weight by iterating on different materials and designs using Finite Element Analysis. It also includes various studies of their manufacturing processes and compares them with additive manufacturing techniques to check the increase in sustainability and the reduction in material wastage. Reducing material wastage and reducing the carbon footprint using additive manufacturing techniques is the main target of the paper. In this work, a gearbox has been designed for a vehicle as an example. The loads have been calculated and the materials also have been selected for the gears and the casing. The materials selected are the materials that are commonly used in the present industry. Finite Element Analysis and optimization have been done to reduce the weight while maintaining a good safety factor at the same time. Sustainability analysis will further be done to compare the amount of carbon footprint by the two processes. From this study, the paper aims to create a process for machine design that takes sustainability, material usage, and carbon footprint into account apart from conventional considerations like weight, strength, size, etc. This will help promote a sustainable environment which is an increasing need. Finally, the work aims to create a process through an example of how to create a sustainable designing and manufacturing process.

Keywords Sustainability · Additive manufacturing · Automobile components · Carbon footprint · ANSYS

A. Jaganaa · V. B. Taruna · K. N. G. L. Reshawantha · G. Rajyalakshmi (✉) · K. Jayakrishna
School of Mechanical Engineering, Vellore Institute of Technology (VIT), Vellore, Tamil Nadu 632014, India
e-mail: rajyalakshmi@vit.ac.in

J. Kandasamy et al. (eds.), *Progress in Sustainable Manufacturing*, Management and Industrial Engineering, https://doi.org/10.1007/978-981-99-0201-9_7

1 Introduction and Literature

In the era of modern technology almost all optimization operations are taking place at a detailed design level that include optimizations of structures of metal gauges as well as beam cross-sections, topology optimization for the casted components, and calibration of engine. Even Though a lot of successes have been seen, optimization is rarely used and also has a lesser impact than expected for most of the researchers. The reason behind the lesser impact may be due to poor connection between CAD and CAE software, accuracy of computer simulations are predictively insufficient, difficulties occurred during automation and computer analysis iterations, larger simulation times, and less availability of data.

At the same time, sustainability has become a great word. Firstly, what is sustainability and what is the need for sustainability? Sustainability mainly refers to the economic value of any product or component which means that the economic value of the product should be the same or higher over a period of time. Sustainability is important because it improves the quality of our lives. It not only benefits us but the environment, if the sustainability of a component is more then the waste produced becomes less and hence more environmentally friendly.

Eckert et al. have performed a study on design of powertrain of electric hydraulic hybrid vehicle. The paper talks about an extensive strategy for optimizing an electric hydraulic mixture vehicle powertrain operating in series to control through the intelligent adaptive-weight hereditary algorithm technique. Factors of the hydraulic drivetrain and the electric framework have been considered advanced. Additionally, a fuzzy-logic regulator, that yields to the stage of the electric engine turn over stop and force that is applied to the siphon that compresses the gatherer, which is moreover thought to be in the detailing of the optimization issue to tune its membership capacities, rules, and weights. The results of the optimization show that electric hydraulic half and half vehicle powertrain structures can be an extremely alluring impetus innovation in regard to both manageable and conservative viewpoints, successfully lessening battery maturing by the utilization of a powerful thickness hydraulic gatherer, which goes about as a pinnacle power cradle unit [1].

Javorski et al. have made a study on gearbox design which is a multi-speed gearbox type and optimization of shifting control is also done so as to minimize the consumption of fuel and losses in mechanical. The authors of the paper focused on optimization of an ICEV drivetrain using the multi-objective optimization techniques, and stuff moving control focusing on the fuel utilization minimization, emanations of exhaust, and power of the gearbox misfortunes. The issue in optimization is addressed by the (I-AWGA) and includes various plan factors of the multi-speed transmission and differential framework, thinking about helpful constraints. The model of the vehicle is assessed with a consolidated driving cycle, accordingly hearty powertrain arrangements may be acquired by the interaction with optimization. The good compromise arrangement brings about the decrease of gas discharges in 2.32% HC, 3.44% CO, and 23.78% NOx, alongside the 15.6% fuel investment funds, confronting the standard vehicle [2].

Haishang et al. have made a work on AM of recycled plastics, many techniques were made for a more sustainable future. In this review, to alleviate any gamble brought about by creation speed, scaling, and speeding up the move towards the more limited recycling and assembling of plastic parts and parts, an assortment recycling-producing (CRM) model is worked to imagine the assessment of cycle stream as well as interaction joining. The review uncovers that AM sets out open doors, for example, prototyping, redoing, transportation cost decrease, and making of occupations in country regions, which might stop superfluous movement; and, above all, diminishing CO_2 discharges and plastic waste regardless of difficulties like abilities prerequisite and detriments in speed and scale creation [3].

Rajak and Vinodh et al. have made a study on application of fuzzy logic for social sustainability performance evaluation: a case study of an Indian automotive component manufacturing organization. This article presents a methodology for social sustainability execution assessment. The methodology has been tried and carried out in an Indian auto part fabricating association. The acquired record has been approved utilizing the traditional fresh method and the social sustainability file is viewed as 6.98. The methodology is productive in estimating the qualities and shortcomings of a singular local area or association concerning a point-by-point set of markers to recognize the more fragile traits. The outcomes utilizing the fuzzy methodology have been approved with the ordinary fresh methodology. 22 social sustainability ascribes out of 60 are viewed as more fragile and fitting activities were inferred to work on the more vulnerable qualities [4].

Sargini et al. have done a study on additive manufacturing of a brake pedal of an automotive vehicle using metal fused deposition modelling. The primary goal of the exploration is the investigation of another brake pedal considering additive manufacturing as an opportunity, combination of parts with decreased mass, new material for additive manufacturing, and utilization of metal-based additive manufacturing innovation. Finite Element Analysis (FEA) has been used to examine how feasible another brake pedal plan for additive manufacturing handling is. The model of the FDM-created metal brake pedal has been actually tried for approving the Finite Element Analysis results and also to check the unwavering quality in FDM innovations that are based out of metal [5].

Mani et al. have made a study on this paper that initially analyses the likely ecological effects of additive manufacturing. A procedure for the sustainability portrayal of additive manufacturing is taken into consideration to fill in an asset for the local area benchmarking additive manufacturing processes in attaining sustainability. The proposed diagram for a sustainability portrayal manual for fill-in as a source of perspective for the local area to benchmark AM processes for sustainability. The aide is still needed to be officially evolved by and by a functioning work thing inside the ASTM E60.13 council [6].

Niaki et al. have focused on reasons behind manufacturers adopting additive manufacturing technologies: The sustainability role in it. This paper aims to recognize and focus on the determinants of its reception and also to explain the benefits of sustainability as the choice to embrace. Thereafter, at that point, the examination tries to recognize the needs of various application areas through a study that is of multiple stages. The outcomes prove that natural sustainability benefits are scarcely applicable to reception choices by and by and this is conversely, with the writing expressing the enormous sustainability benefits. The outcomes demonstrate the significant job of financial thought processes in reception choices. The discoveries additionally demonstrate that the capacity of added substance fabricating for delivering practically any perplexing plan is the vital driver of its reception in all areas [7].

Wie et al. have made research on Failure analysis for a transmission gear used in an electric vehicle which is known as secondary driving helical gear. In this paper, the observed results were disappointing due to auxiliary driving helical stuff in the transmission arrangement of an EV being found to be dissected. The pressure conveyances of the tooth flank taken utilizing dynamic and static contact in finite elements was acquired again in view of ANSYS Workbench programming. The outcomes then showed that the surfaces that were cracked in the bombed gear begin from the roots of tooth and display the small granular fragile break, the fundamental explanation, solidifying and hardness profounding in the root area, various huge carbides on the network of martensite because of ill-advised hotness treatment, additionally FE reproduction results uncovered that more contact pressure in the root position and high effect force is acting in the underlying phase of cog wheels fitting while minimizing in assistance perhaps prompted the weakness disappointment [8]. Qingyong et al. have focused on contact mesh analysis and topology optimization of electric vehicle gearbox. Taking into account the NVH issue of car transmissions, the third pair of stuff sets of an electric vehicle gearbox is taken as the examination object. In view of the nonlinear unique model hypothesis of the stuff transmission framework, the stuff profile alteration and tooth direction are exhaustively used to choose a sensible shape change plot. Noticing the reproduction results, it tends to be seen that the stuff pair transmission blunder and the most extreme contact pressure after the shape alteration are altogether diminished contrasted and those before the shape change, and the contact pressure map circulation is more uniform [9]. Srikar and Mahato have made a study on design of two stage single speed gearbox used in terrain vehicles along with calculation and its analysis. In this paper, they zeroed in on the hypothetical investigation of planning and estimation, along with its analysis of gearbox which can be used in a BAJA SAE. The gearbox is further co-ordinated by CVT, i.e., constant factor transmission which is combined with ultraviolet joints for transmission of force to the wheels. The gearbox is laboured for 2 years and is considered with productive resilience [10].

Feucht et al. have analysed how 3D printing can be used for components made with steel in additive manufacturing. They focused mainly on preliminary strength investigations that were carried out during the process, and also the sequence of processes that are required for homogeneous manufacturing. Author's focus is especially around the starter strength examinations completed and the interaction groupings expected for homogenous assembling. The paper finishes up with a thought of the execution on location [11].

In the present world reducing the weight of an automotive component is of great requirement without getting induced stresses to it. Reduction of body weight of automotive components also helps in getting a good mileage for a car or a bike. The authors

M. R. Idris, S. A. Syed Ahmed, E. Sujatmika, and W. M. Wan Muhamad, showed that 24% mass reduction can be done for the design of the rear spindle. In the same way we can try for other vehicle components which will help us contribute towards environmental sustainability, better conserving the world's metal resources and reducing carbon emission through improved overall vehicle fuel efficiency [12]. Considering additive manufacturing techniques such as 3D printing and CNC machining the production of automotive components becomes much easier with less wastage, possibility of producing every form and function. But the authors Enrico Dalpadulo, Fabio Pini, and Francesco Leali gave a conclusion that there is no added advantage in replacing the integrated platform with a stand-alone tool, rather than implementing the whole method and the workflow into a platform [13].

As common people these days are aware of buying sustainable products, the plastic industry is doing drastic changes in their manufacturing processes by considering 3D printing and other techniques of production in the industry. They also developed a methodology that demonstrates the advantage in comparing the existing subjective optimization research processes [14]. In order to analyse the sustainability in industry there are so many factors and variables needed. Some of the variables that are generally considered are consumption of energy, consumption of water, waste management, environment preservation, equality in society, and noise and emission management. Waste that is generally generated in the automotive manufacturing industry are classified as machine lubricants, coolants, Solvent cleaning, paint and scrap metals, and plastics. The authors C. Torcătoru and D. Săvescu stated that the above-mentioned variables accounted for a lot of variation in sustainability and also showed that an 83% variation in sustainability can be achieved which is shown using the statistics based calculation such as using the multiple regression, etc. [15].

It can also be used as a list of KPIs proposed by the authors Vikas Swarnakar, A. R. Singh, and Anil Kr Tiwari in their paper which is based on AHP (Analytical Hierarchy Process) in order to know the expert's opinion. They also proposed that few of the sustainability parameters are economic, social, and environmental. And the KPIs that come under economic are operational cost, the rate of acceptance of the product, effectiveness of overall equipment, inventory level work process, performance of equipment and machine, cost of facilities, efficiency in transportation; KPIs under social are satisfaction of employees, relation with the labour, society contribution, rate of accidents, opportunities in training, absenteeism ratio, volunteer sustainability initiatives, and gender ratio; and KPIs under environmental conditions are toxic water discharge rate, impact on green area, releasing harmful gases, consumption of water and fuel and also the materials, and overall solid waste generation [16].

A case study is presented by C. Torcătoru and D. Săvescu in which the authors chose a product which is analysed with SOLIDWORKS Sustainability and they also generated the sustainability report provided by SOLIDWORKS which is saved and used for the adoption of the right decision by the whole organization [17].

Process optimization parameters that are considered for manufacturing a spur gear as these fears are of different types which are most likely to be used in all types of power transmission systems. The design is further analysed theoretically and through finite element analysis and both the theoretical and practical results were compared.

The comparison showed that the theoretical results were far better than the finite element analysis results [18].

The authors have done AIP Conference Proceedings. Vol. 2347. No. 1. AIP Publishing LLC, 202 re-designing a bluetooth speaker which is more compact and sleeker looking and it should also be a water-resistant bluetooth speaker. For the purpose of designing, DFMA is used for minimization of quality required, and SOLIDWORKS Sustainability is used for evaluating for comparing different materials with four environmental factors such as carbon footprint, water eutrophication, total energy consumption, air acidity, and as well as material financial impact too [19, 20]. Jones, F. tried generating a simulation tool which is capable of predicting properties of heat-treated by the processes called carburizing-quenching-tempering, and finally display the results to the user in an understandable manner of comparison between real and simulation-based results of surface hardening of automotive components have been done [21].

After going through the various research papers, a lot has been learnt regarding the area of study. Literature regarding optimization of different components has been studied which mainly includes gearbox optimization. Structural analysis has been learnt from various research papers. The various parameters of gears such as the gear ratio, number of teeth, helical angle, face width, etc., can be iterated to get the best combination of strength and weight. The loads have been applied using static structure in Ansys and the various iterations were carried out.

Various additive manufacturing processes have been studied from the literature. It has been seen how the amount of material used can be reduced by additive manufacturing. Furthermore, sustainability has been studied to utilize the various resources in a judicious manner. The carbon footprint and power consumption of the various processes have been calculated to select the best process. Even though many researchers worked for manufacturing practices for various automobile components, sustainability of manufacturing practices was not reported. The major focus of this work is to optimize the design of automobile components and calculate the sustainability of manufacturing through material reduction, energy consumption, CO_2 emissions, etc. This work also concentrated on comparison of present manufacturing processes with additive manufacturing.

The objective of the paper is to introduce a process for design and optimization that also considers additive manufacturing and sustainability. The project introduces this process by using an example of a four-wheel drive gearbox. Design of gearbox includes calculation of loads, material selection, and optimization. This work is focused on sustainability analysis to compare the carbon footprint of various manufacturing processes.

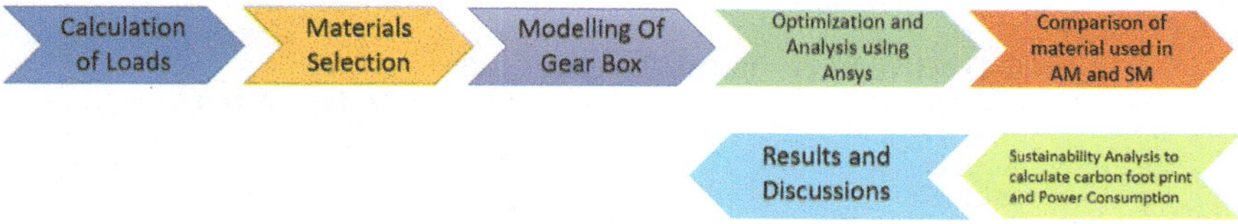

Fig. 1 Methodology of the project

2 Methodology

The methodology starts with the calculation of the loads acting on the different components of the gearbox. The materials then have been selected for the components. The design of the gearbox is then made using SOLIDWORKS. The components have then been analysed using Static Structural in ANSYS. The optimization process was hence completed by performing multiple iterations of the component designs such that the weight is minimum with an appreciable factor of safety (Fig. 1).

The material usage is then calculated in the case of additive and subtractive manufacturing and compared. The amount of material saved by the use of additive manufacturing is then calculated. Sustainability Analysis is further done. The carbon footprint and power consumption of various additive manufacturing processes is then calculated to select the most suitable process. A design process is hence proposed that also takes into account additive manufacturing and sustainability.

3 Design Specifications of Gears

The parameters considered to design a gear are given in the following Table 1.

Step 1: Calculation of Loads and Material Selection of Components

The total tractive force required for an ATV to climb an incline has been calculated. The gear ratio is calculated from that (Fig. 2).

Table 1 Optimized design parameters of gearbox

S. No	Parameter	Value
1	Tractive force	2292.46 N
2	Tractive torque	669.4 Nm
3	Gearbox ratio	9:1
4	Face width	

Gear Design

Total Tractive force= mg Sinθ + μmg Cosθ

= 230*9.81*Sin 60 + 0.3*9.81*Cos 60

= **2292.46** N

Total Tractive Torque= Total Tractive force* Radius

= 2292.46 * 0.292

= **669.4** Nm

Input Torque* Gearbox Ratio= 669.4 N

74 * R = 669.4 N

R=9:1

Face width Calculation, First stage

Z = 15

mn = 3 mm

mt = 3/ Cos 15

= 3.11 mm

d = z* mt

= 15 * 3.11

= 46.65 mm

V = pi*d*n / 60

= (3.14 * 46.65 * 3800) / (60 * 1000 * 0.9)

= 10.31 m/s

Ft = W/V

= 6714/10.31

= 651.21 N

Kv = (Sqrt (18+sqrt (200)/78)

= 1.26

$\sigma e = \sigma e' * k_L\ k_V\ k_S\ k_R\ k_T\ k_F\ k_N$

$\sigma b = \sigma e/2 = 173.34$ Mpa

J = 0.4 * 0.98 = 0.392

$\sigma b = (F_t / b\ M_n\ J) * K_V\ K_D * (0.93 K_m)$

b = 13 mm

Bearing Force, First Stage

V= π*d*N/ 60

= (3.14 * 15 * 3.11 * 20000) / (60 * 1000 * 3.9)

= 1.25 m/s

F_T = W/V

= 4000 / 1.25

= 3200 N

$F_M = F_T \tan \varphi / \cos \beta$

= 3200 tan 20 / Cos 15

= 1205.79 N

$F_a = F_T \tan \beta$

= 3200 * tan 15

= 857.44 N

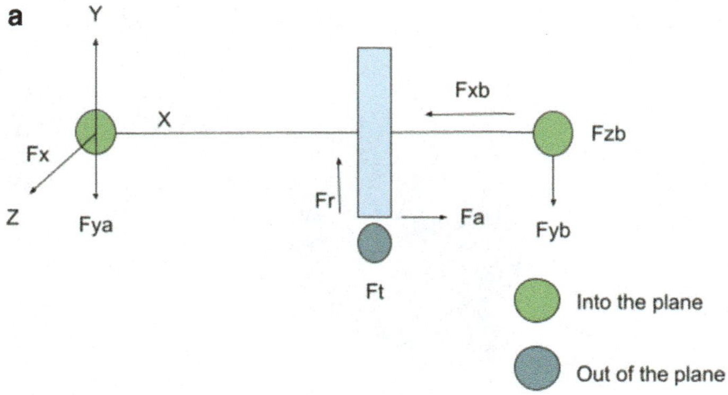

Fig. 2 **a** Design dimensions with calculations. **b** Design dimensions with calculations

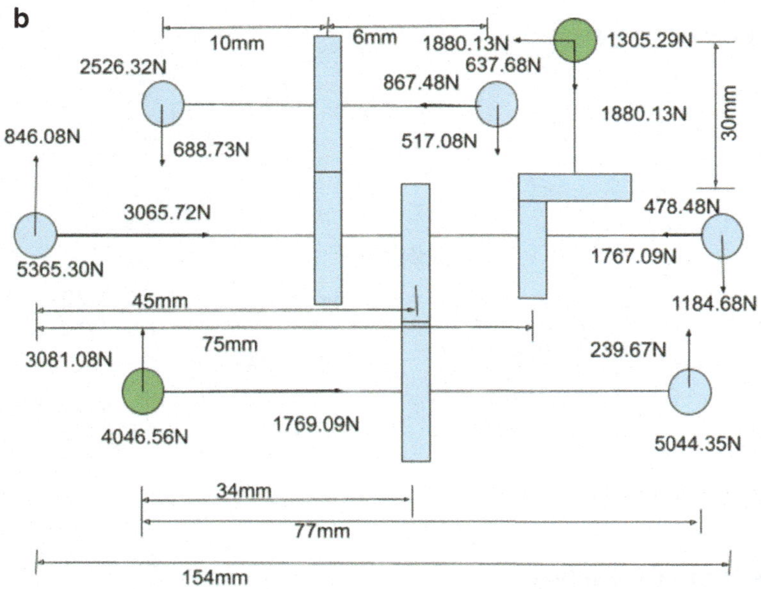

Fig. 2 (continued)

Gear 1

$$(\text{TE (z- axis)})\ Fyb * 76 - Fx * 16 - Fa * 23.33 = 0$$
$$Fyb = 517.06\ \text{N}$$
$$(\text{FE (y- axis)})\ Fya - Fr + Fyb = 0$$
$$Fya = 688.73\ \text{N}$$
$$(\text{TE (y- axis)})\ Ft * 16 = Fzb * 76$$
$$Fzb = 673.68\ \text{N}$$
$$(\text{FE (z- axis)})\ Fza - Ft + Fzb = 0$$
$$Fza = 2526.32\ \text{N}$$

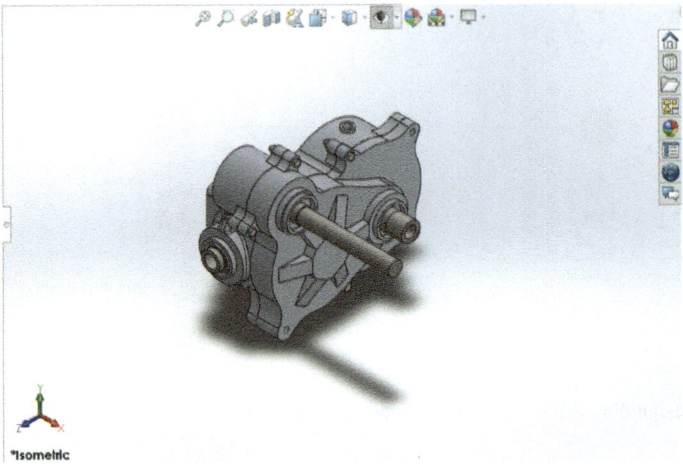

Fig. 3 Design of the gearbox

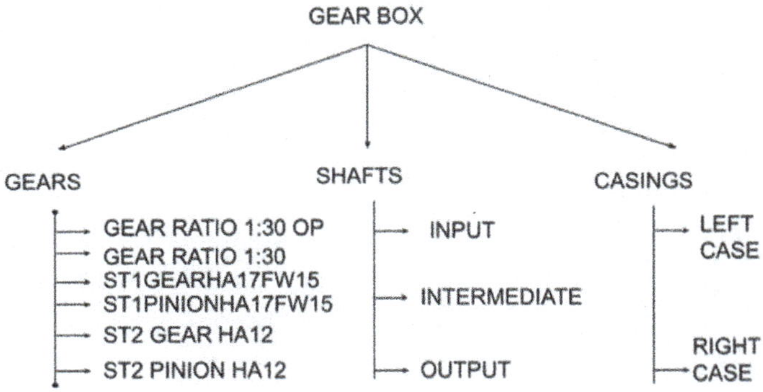

Fig. 4 Components included in a gearbox

4 Design of Gearbox

A gearbox has been designed in SOLIDWORKS considering the various design parameters studied in various research papers (Figs. 3 and 4).

5 Structural Analysis and Optimization of the Components

Structural analysis has been done on the components. The loads calculated have been applied. The simulation has been done using Ansys 2020 R2. Weight has been reduced while keeping an appreciable safety factor (Fig. 5).

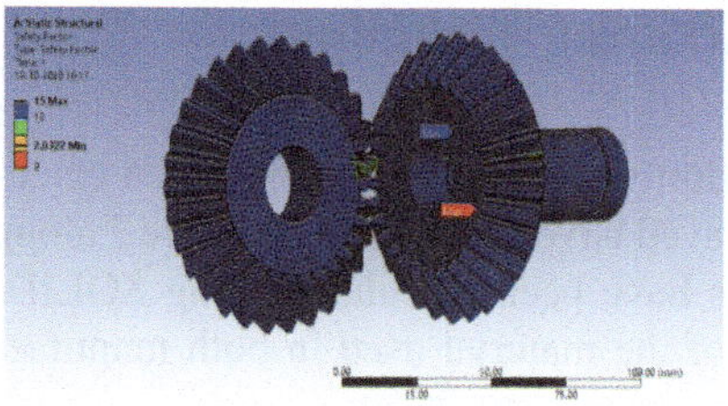

FOS: 2.0322

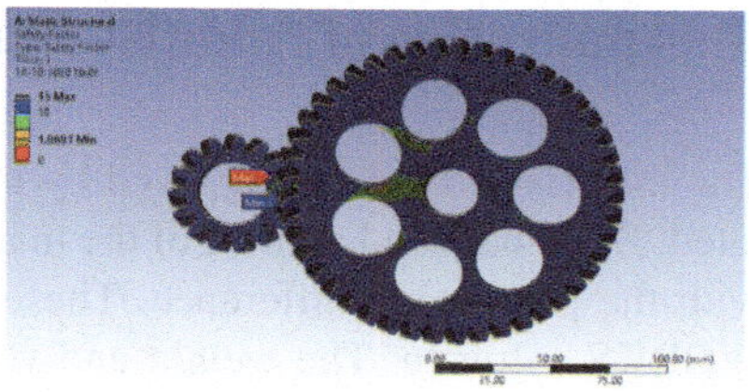

FOS: 1.9697

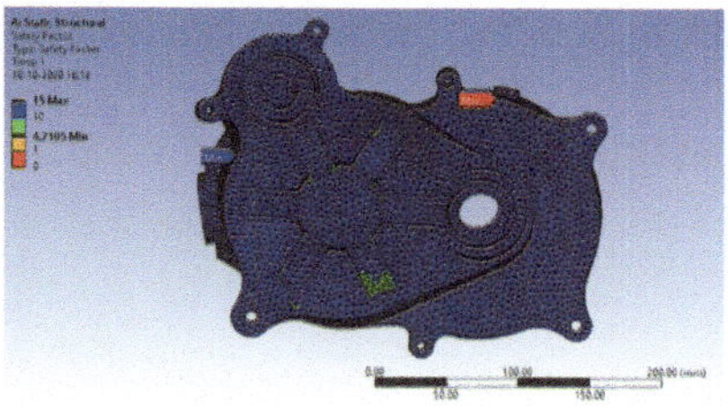

FOS:4.2105

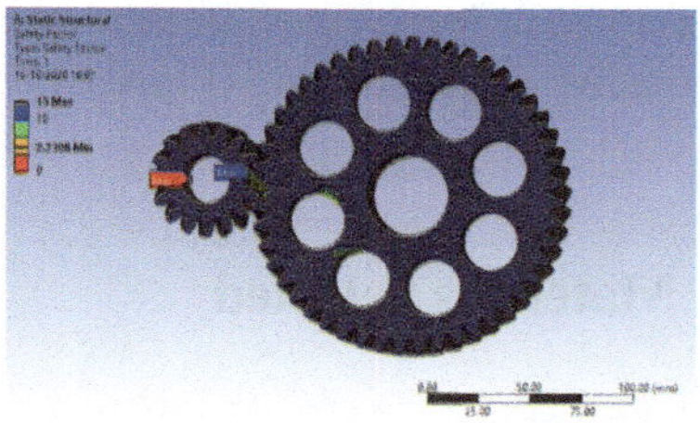

FOS:2.2308

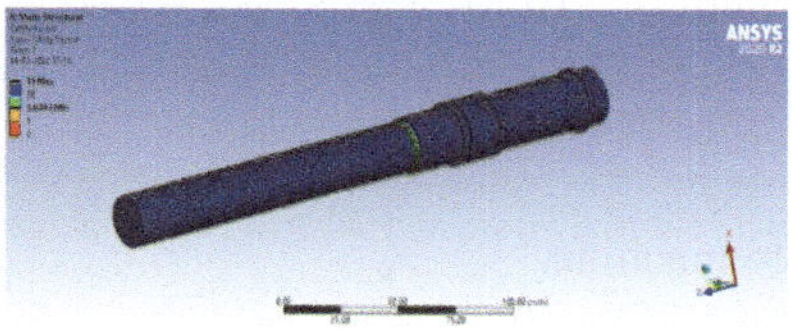

FOS: 5.0493

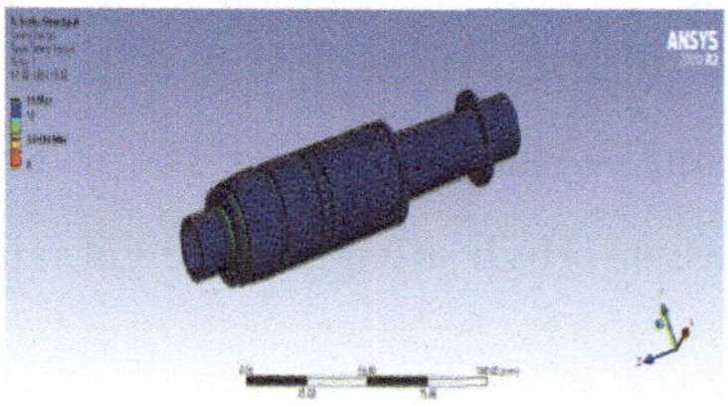

FOS: 3.0414

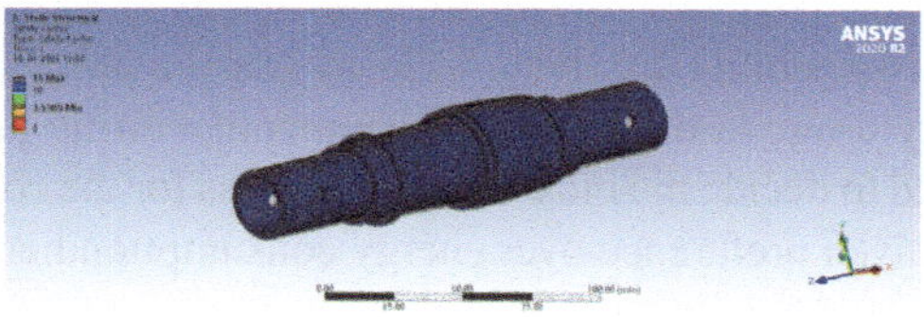

FOS: 3.5589

Fig. 5 Analysis and optimization of the component

6 Application of Additive Manufacturing Techniques

The material that has been saved by using additive manufacturing has been calculated. Both mass and volume of the material saved have been shown in Table 2 along with the percentage difference. These values have been calculated using SOLIDWORKS software. The weight and volume of the material used in both manufacturing processes, i.e., subtractive and additive manufacturing processes have been calculated and compared.

The following images illustrate the designs in both the cases, i.e., additive and subtractive manufacturing (Fig. 6).

7 Materials Selected

Gears—AISI 4340

Reason—Ease of machining, High Yield Strength of 710 MPa which is 16000 psi more than that of AISI 4130 (Fig. 7).

Casing—Aluminium 6061 T6.

Reason—High strength to weight ratio, Easy to machine.

8 Sustainability Analysis

SOLIDWORKS Sustainability has been used to analyse and compare the carbon footprint and energy usage in different additive manufacturing processes. The data has further been used to decide the most suitable process for the components. Carbon footprint has been given preference over energy consumption here.

8.1 Input Parameters Values for Manufacturing of Gearbox

For sustainability in SOLIDWORKS, prior one needs to consider a few parameters such as what type of material we are going to use for our product, what is the location of manufacturing, where the product will be used, and how many years it is built to last. For this the following table shows the parameters considered for this project (Table 3).

Table 2 Material usage comparison between additive and subtractive manufacturing process

	Subtractive		Additive					
	Volume (cm^3)	Mass (g)	Volume (cm^3)	Mass (g)	Volume difference	Mass difference	Vol %	Mass %
Gears								
First stage								
Driving gear	35,637.4417	279.7539	17,590.9083	138.0886	18,046.5334	141.6653	50.63925057	50.63925829
Driven gear	252,640.5713	1983.2285	97,271.1427	763.5785	155,369.4286	1219.65	61.49820981	61.4982086
Second stage								
Driving gear	73,844.6047	579.6801	47,624.5802	373.853	26,220.0245	205.8271	35.50702804	35.50701499
Driven gear	571,565.1532	4486.7865	228,809.8723	1796.1575	342,755.2809	2690.629	59.96784076	59.96784113
Bevel stage								
Driving gear	128,816.188	1011.207	66,340.984	520.777	62,475.204	490.43	48.49949759	48.49946648
Driven gear	573,862.753	4504.823	97,731.483	767.192	476,131.27	3737.631	82.96953714	82.96954176
Shafts								
Input shaft	166,715.21	1308.71	109,707.79	861.21	57,007.42	447.5	34.19449251	34.19397728
Intermediate shaft	93,902.2	737.13	42,062.32	330.19	51,839.88	406.94	55.2062465	55.20600166
Output shaft	248,311.48	1949.25	119,828.23	940.65	128,483.25	1008.6	51.74277484	51.74297807
Casing	10,828,374.44	29,236.61	1,102,183.52	2975.9	9,726,190.92	26,260.71	89.82133906	89.82132333
				With Al	11,044,519.21	36,609.5824		
				Without Al	1,318,328.291	10,348.8724		

a

Fig. 6 **a** Material usage comparison of first stage gears. **b** Material usage comparison of second stage gears. **c** Material usage comparison of bevel gears. **d** Material usage comparison of casing. **e** Material usage comparison of shafts

8.2 Carbon Footprint

While burning the fossil fuels gases such as carbon dioxide and many others add up into the atmosphere resulting in an increase in the earth's temperature. Global warming potential is the impact factor used for the carbon footprint value of the earth. Global warming also causes problems like species extinction, evaporation of water bodies, severe weather conditions, etc. Tables 4, 5, 6, and 7 show carbon footprint values for different components such as gears, shafts, and casings of the gearbox which are calculated in SOLIDWORKS for different manufacturing processes like extrusion, die casting, sand casting, and machined sand casting.

Fig. 6 (continued)

8.3 Energy Consumption Values for Manufacturing of Gearbox

Non-renewable energy resource consumption is calculated in megajoules (MJ). As a result of this consumption of electricity and fuels during the product manufacturing some or the other people on earth would be suffering due to their dependence on these energy resources. The average calorific value of energy demand of the non-renewable resources is expressed as total energy consumed. Tables 8, 9, 10, and 11 show energy consumption values for different components such as gears, shafts, and casings of the gearbox which is calculated in SOLIDWORKS for different manufacturing processes like extrusion, die casting, sand casting, and machined sand casting.

The results mentioned in the tables show us that the carbon footprint values for gears in the extrusion process is less when compared to die casting, sand casting,

c

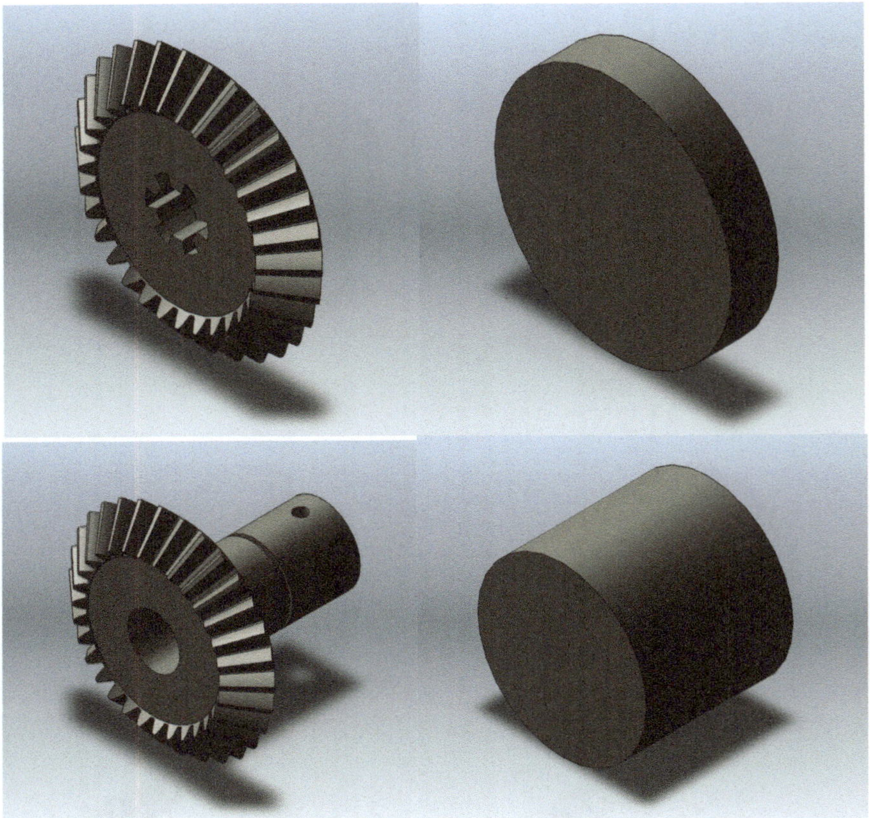

Fig. 6 (continued)

d

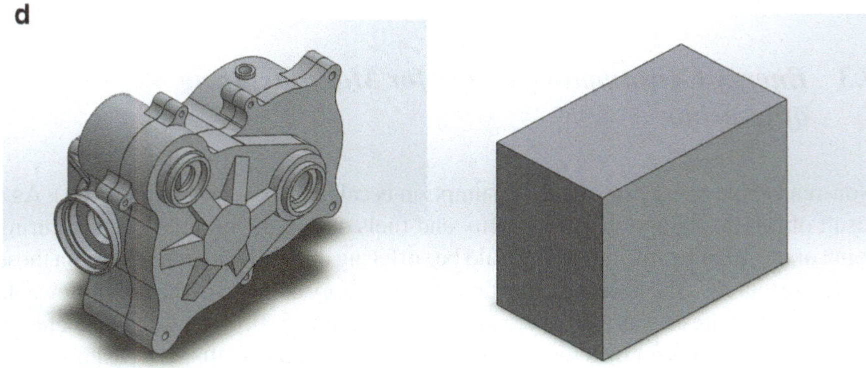

Fig. 6 (continued)

Fig. 6 (continued)

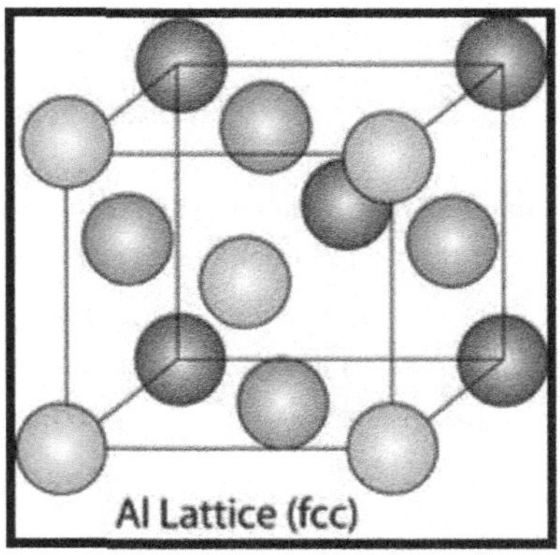

Fig. 7 Lattice structure of Aluminium 6061 T6

Table 3 Input parameters for sustainability analysis

Material class	Aluminium alloys
Material name	AISI 4340 for gears and shafts and for casing 6061-T6(SS)
Built to last	2 years
Region of manufacturing	Asia
Region of use	Asia
Manufacturing processes	Extrusion, die casting, machined sand casting, and sand casting

Table 4 Carbon footprint comparison of all gears in different manufacturing processes

	Extrusion	Die casting	Sand casting	Machined sand casting
Gear1	0.19	2.7	2.7	3.2
Gear2	0.141	2	2	2.4
Gear3	0.198	2.8	2.8	3.3
Gear4	0.032	0.453	0.453	0.536
Gear5	0.446	6.3	6.3	7.4
Gear6	0.089	1.3	1.3	1.5
	All the values are in Kg CO_2			

Table 5 Carbon footprint comparison of all shafts in different manufacturing processes

	Extrusion	Die casting	Sand casting	Machined sand casting
Shaft1	0.228	3.2	3.2	3.8
Shaft2	0.096	1.4	1.4	1.6
Shaft3	0.244	3.4	3.4	4.1
	All the values are in Kg CO_2			

Table 6 Carbon footprint comparison of all casings in different manufacturing processes

	Extrusion	Die casting	Sand casting	Machined sand casting
Casing1	1.5	2.4	1.4	1.7
Casing2	1.9	3.1	1.8	2.2
	All the values are in Kg CO_2			

Table 7 Average carbon footprint values of gears, shafts, and casings

	Extrusion	Die casting	Sand casting	Machined sand casting
Gears	0.18267	2.592167	2.592167	3.056
Shafts	0.18933	2.6667	2.6667	3.1667
Casings	1.7	2.75	1.6	1.95
	All the values are in Kg CO_2			

Table 8 Energy consumption comparison of all gears in different manufacturing processes

	Extrusion	die Casting	Sand casting	Machined sand casting
Gear1	16	24	15	18
Gear2	1.7	21	21	25
Gear3	2.3	30	30	35
Gear4	0.38	4.8	4.8	5.7
Gear5	5.3	67	67	79
Gear6	1.1	13	13	16
	All the values are in MJ			

Table 9 Energy consumption comparison of all shafts in different manufacturing processes

	Extrusion	Die casting	Sand casting	Machined sand casting
Shaft1	2.7	34	34	40
Shaft2	1.1	14	14	17
Shaft3	2.9	37	37	43
	All the values are in MJ			

Table 10 Energy consumption comparison of all casings in different manufacturing processes

	Extrusion	die casting	Sand casting	Machined sand casting
Casing1	16	24	15	18
Casing2	21	31	19	23
	All the values are in MJ			

Table 11 Average energy consumption values of gears, shafts, and casings

	Extrusion	Die casting	Sand casting	Machined sand casting
Gears	4.4633	26.633	25.133	29.7833
Shafts	2.233	28.333	28.333	33.333
Casings	18.5	27.5	17	20.5
	All the values are in MJ			

and machined sand casting and hence extrusion is the best additive manufacturing process for manufacturing of gears. In a similar fashion for shafts, extrusion is the best and for the casing, sand casting is the most suitable additive manufacturing process.

9 Results and Discussions

The design of the component has been completed using SOLIDWORKS. It has been analysed using Ansys and further optimized. An appreciable safety factor has been kept for all the components. The components' material usage has then been compared between both additive manufacturing and subtractive manufacturing scenarios. In this case, approximately 56.51% of the material has been saved in the case of additive manufacturing.

Further analysis is done to find out which additive manufacturing process can be used in order to have the least carbon footprint. SOLIDWORKS Sustainability has been used for this purpose. In the case of gears and shafts that are made of steel, extrusion comes out to have the least carbon footprint of 0.18267 KgCO_2 and 0.18933 KgCO_2, respectively. Sand Casting comes out to have the least carbon footprint of 1.6 KgCO_2 in the case of the casings that are made of aluminium.

The energy consumption has also been analysed. In this case, the results obtained are different. Gears and casings have the least energy consumption of 4.4633MJ and 2.233 MJ, respectively, in the case of the sand casting process. Manufacturing of shafts takes the least energy consumption of 17MJ when done through the extrusion process.

10 Conclusion

An optimized design of the gearbox has been completed using SOLIDWORKS and Ansys. The weight of the components has been minimized while maintaining an appreciable safety factor. Further, the material used has been compared between additive and subtractive manufacturing. It has been found that additive manufacturing results in the saving of approximately 56.51% of the material. It is hence better in this aspect. The sustainability analysis has further been done where the carbon footprint and energy consumption in various additive manufacturing processes have been compared to find the optimum process for manufacturing. The carbon footprint is the total amount of greenhouse gases generated in the processes. The power consumption talks about the total amount of electricity consumed during the process. The optimum process would then be decided by checking the values of these factors. If carbon footprint is to be given priority, then the process with the least carbon footprint would be preferred. If power consumption is to be given priority, then the process with the least power consumption would be preferred.

A new design process has hence been proposed which would also take into account the amount of material used along with the sustainability. The process will take into account important sustainability factors such as carbon footprint and power consumption.

References

1. Eckert JJ et al (2022) Electric hydraulic hybrid vehicle powertrain design and optimization-based power distribution control to extend the driving range and battery life cycle. Energy Convers Manag 252
2. Eckert JJ et al (2022) Multi-speed gearbox design and shifting control optimization to minimize fuel consumption, exhaust emissions and drivetrain mechanical losses. Mechan Mach Theory 169:104644
3. Wu H et al (2021)Additive manufacturing of recycled plastics: Strategies towards a more sustainable future. J Clean Prod 130236
4. Rajak S, Vinodh S (2015) Application of fuzzy logic for social sustainability performance evaluation: a case study of an Indian automotive component manufacturing organization. J Clean Prod 108:1184–1192
5. Sargini MIM et al (2021) Additive manufacturing of an automotive brake pedal by metal fused deposition modelling. Mater Today: Proc 45:4601–4605
6. Mani M, Lyons KW, Gupta SK (2014) Sustainability characterization for additive manufacturing. J Res Natl Inst Stand Technol 119:419
7. Niaki MK, Torabi SA, Nonino F (2019) Why manufacturers adopt additive manufacturing technologies: the role of sustainability. J Clean Prod 222:381–392
8. Feng W, Feng Z, Mao L (2020) Failure analysis of a secondary driving helical gear in transmission of electric vehicle. Eng Fail Anal 117:104934
9. Zhang Q et al (2020) Contact mechanics analysis and optimization of shape modification of electric vehicle gearbox. Jordan J Mech Ind Eng 14(1)
10. Reddy BS, Mahato KK (2021) Calculation, design and analysis of two stage single speed gearbox for all terrain vehicle for Baja sae. Mater Today: Proc 46:7187–7203
11. Feucht T et al (2021) 3D-printing with steel: additive manufacturing of a bridge in situ. Ce/ Pap 4(2–4):1695–1701
12. Muhamad WMW et al (2012) An optimization analysis on an automotive component with fatigue constraint using hyperworks software for environmental sustainability. Int J Mech Mechatron Eng 6(8):1395–1399
13. Dalpadulo E, Pini F, Leali F (2020) Integrated CAD platform approach for design for additive manufacturing of high performance automotive components. Int J Interact Des Manuf (IJIDeM) 14(3):899–909
14. Romero VJ, Sanchez-Lite A, Liraut G (2022) Development of a multi-criteria design optimization methodology for automotive plastics parts. Polymers 14(1):156
15. Juniarty M, Ismail Y (2015) Analyzing sustainability of SMEs in automotive component industry in Bekasi regency. J Asian Sci Res 5(12):522–533
16. Swarnakar V, Singh AR, Tiwari AK (2021) Evaluation of key performance indicators for sustainability assessment in automotive component manufacturing organization. Mater Today: Proc 47:5755–5759
17. Torcătoru C, Săvescu D (2019) Analyzing the sustainability of an automotive component using SolidWorks CAD software. IOP Conf Ser: Mater Sci Eng 568(1) (IOP Publishing)
18. Apparao D, Raju MVJ (2021) Design and analysis of spur gear manufactured by DMLS process. Mater Today: Proc 46:149–153
19. Altıparmak SC et al (2021) Challenges in additive manufacturing of high-strength aluminium alloys and current developments in hybrid additive manufacturing. Int J Lightweight Mater Manuf 4(2):246–261
20. Effendi MSM et al (2021) Integrating DFMA to sustainability and FEA assessment: Case study of portable Bluetooth speaker. AIP Conf Proc 2347(1) (AIP Publishing LLC)
21. Jones F (2021) Finite element analysis of surface hardening treatments of steels for automotive powertrain applications. University of Windsor (Canada), Diss

A Brief Review of Sustainable Composites for Food Packaging Applications

Soundhar Arumugam and Senthilvelan Selvaraj

Abstract The food industry is currently confronting a wide range of difficulties in ensuring the quality of food with a longer shelf life and long-term preservation. The industries have been able to tackle these issues due to sustainable and biodegradable food packaging materials based on biopolymers. Due to their superior biodegradability, renewability, bioavailability, and non-toxicity, these eco-friendly materials are also reducing the environmental issues connected to plastic-related pollution. Presented here are biopolymer-based food packaging materials and their composites that influence the qualities of food packaging. This analysis also highlights existing research studies for leveraging these ecologically friendly materials in food packaging applications.

Keywords Biopolymers · Sustainable materials · Food packaging · Biodegradable materials

1 Introduction

Food preservation is greatly aided by food packaging, which surrounds the food in a barrier of protection. Food packaging not only contributes to the preservation of product quality and safety but also to the extension of the product's packaging life against any physical, chemical, biological, or environmental issues during storage and shipment [1–3]. Microbial contaminations increase the risk of foodborne infections, shorten the shelf life of food, and cause food products to deteriorate [4, 5]. Economic losses due to food spoiling totaled more than 25% of the food consumed [6]. Researchers have made valiant attempts to create novel food preservation techniques to address these problems. Nowadays, the addition of various food additives aids in the long-term preservation of food by preventing microbial deterioration [7, 8]. Foods are protected for a long time by properly designed active agent carriers

S. Arumugam · S. Selvaraj (✉)
Department of Mechanical Engineering, Indian Institute of Technology Guwahati (IITG), Assam 781039, India
e-mail: ssvelan@iitg.ac.in

J. Kandasamy et al. (eds.), *Progress in Sustainable Manufacturing*, Management and Industrial Engineering, https://doi.org/10.1007/978-981-99-0201-9_8

because they release their agents under controlled conditions when they come into contact with food [9]. These carriers are also appropriate substitutes for conventional preservation methods that allow food to absorb active ingredients without altering its organoleptic qualities, such as appearance, colour, flavour, aroma, and taste [10]. Nearly 85% of plastic garbage ends up in the ocean, significantly affecting both the ecology and wildlife [11, 12]. Furthermore, the consumption of synthetic polymer is continuously growing with time. According to a report, by the year 2050, there will be more than 25 billion metric tonnes of non-biodegradable plastic garbage in the world. Additionally, estimates show that approximately 8 billion tonnes of plastic have been manufactured during the past 70 years [13–15].

The "triple P"—profit, planet, and people are all balanced by sustainability. Researchers are constantly looking for renewable and bioresources because of the various "P"-related factors, such as a limited supply of petroleum-based products and an environmentally friendly atmosphere [16]. The greatest materials for creating innovative sustainable solutions are those based on biopolymers. Due to this, the world is exploring replacing petroleum-based plastics with sustainable biodegradable polymers, with or without the addition of natural fibres. Currently, various biopolymers used in food packaging applications such as polyvinyl alcohol (PVA), polylactic acid (PLA), polyglycolic acid (PGA), polybutylene succinate (PBS), polycaprolactone (PCL), Polyhydroxybutyrate (PHB), chitosan, cellulose, agar, and starch [7, 17, 18]. In today's world, there are numerous opportunities for industrial sectors to progress towards a cleaner, more sustainable, and greener environment. Consuming various biopolymers can help reach these long-term objectives since they have a variety of benefits, such as abundant availability in nature, reasonable mechanical qualities, and ease of biodegradation [19]. Various biopolymers currently support the world's quickly expanding markets. By 2025, all major plastic packaging manufacturers are anticipated to begin manufacturing packing materials that are fully recyclable, biodegradable, and reusable [15]. However, when compared to the petroleum-based materials, the use of biopolymers as food packaging materials has limitations such as weaker mechanical, thermal, and barrier properties. As a result, various research efforts have been made to improve the properties of biopolymers. In order to improve the biopolymer properties, nanocomposite concept was developed [20]. Bio-nanocomposites are multiphase materials composed of continuous phase of matrix particularly biopolymer and discontinuous (reinforcement) phase of nanofillers. The nano-sized fillers play a significant role by acting as reinforcement to enhance the mechanical and barrier properties of the biopolymer [21]. The incorporation of lignocellulosic fibres such as wheat straw, flax, sisal, jute, kenaf, coconut shell, grape pomace, olive pomace, almond shell, and rice husk into biopolymers improves the mechanical and barrier properties and also provides antimicrobial agent, biosensor, and oxygen scavenger functions in food packaging application [22]. Furthermore, biopolymers and their composites have prospective uses in a variety of industries, including medication delivery, packaging, textiles, agriculture, disposable materials, and automotive applications [23]. The life cycle of biodegradable materials is illustrated in Fig. 1.

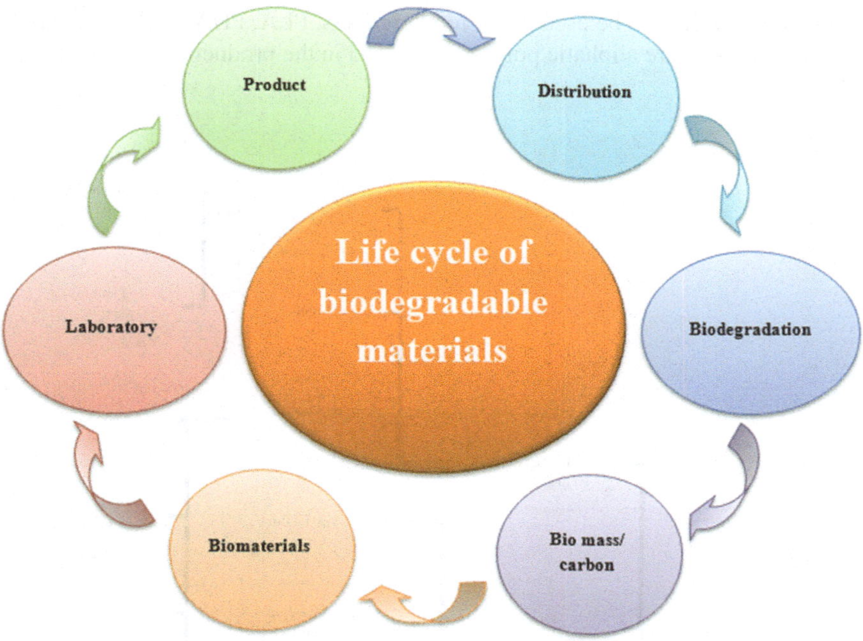

Fig. 1 Life cycle of biodegradable materials

2 Biopolymers

Different factors, such as moisture content, temperature, and rancidity, all play a vital part in preserving quality of the food. In recent years, biopolymer-based materials, a subclass of bioplastics and biodegradable plastics, have fascinated the food packaging market [24]. Biopolymer-based packaging materials are derived naturally from lipids, proteins, aliphatic polyesters, or polysaccharides and are being evaluated as possible candidates for conventional plastics due to their biocompatibility, safety, and rate of biodegradation. Natural micro-organisms efficiently break down these polymers in the presence of adequate temperature, moisture, and oxygen, without any environmental consequences [25]. Biopolymers can be polymers extracted directly from biomass of vegetable or animal origin, such as polysaccharides, proteins, and lipids, as well as polymers developed by classical chemical synthesis from renewable bio-based monomers, such as poly (lactic acid) (PLA), or polymers produced by wild or genetically modified micro-organisms, such as poly(hydroxyalkanoate)s (PHAs), poly(hydroxybutyrate)s (PHBs), bacterial cellulose, and xanthan [26].

Figure 2 shows the classification of biopolymers used in packaging materials. Starch, cellulose, chitosan, alginate and carrageenan, pectin, and gums or their derivatives are examples of polysaccharides. Protein films were classified as either animal (gelatin, casein, whey, and collagen) or vegetable (zein, soy, and gluten) [27, 28]. Beeswax and carnauba wax are the most commonly utilised lipids in the production of emulsion coatings to be used in the food products. PLA, PHA, PHB, and various polyamide classes are aliphatic polyesters utilised in the production of biopolymers [10].

Fig. 2 Classification of biopolymers

2.1 Biopolymer-Based Nanocomposites

Traditional packaging technology employs non-biodegradable polymer composites, which have serious environmental consequences for human health and land fertility by slowing the rate of oxygen passage in the soil [29]. In order to address this issue, biopolymeric sustainable composites are being used in the food service industry. These composites are created by combining two or more biopolymeric components, either with or without the use of additives/NFs. These composites have demonstrated good biodegradability and have been studied by a number of researchers [13]. Blending biopolymers increases mechanical and functional qualities, which are useful in applications like food packaging. Dimensions, shape, length, surface area, and aspect ratio of nano-additives are all influencing elements in the attributes of nano-biocomposites. Furthermore, the permeability of these biodegradable composites is determined by the orientation and dispersion of the nanoparticle (NPs), resulting in strong metallurgical bonding between the NPs and the polymer matrix [30]. Sharma et al. [31] prepared the composite film using polylactic acid (PLA) with halloysite nanotubes and ramie fabric by melt extrusion method. The results displayed that the relative loss in tensile strength of ramie-PLA composites was more compared to neat PLA. Additionally, ramie-PLA composite has completely broken down after 60 days in the soil [31]. Manikandan et al. 2020 prepared nanocomposite films made up of Polyhydroxybutyrate (PHB) reinforced with various concentration of (0–1.3 wt.%) of graphene nanoplatelets (Gr-NPs) using solution casting method. The PHB/Gr-NPs nanocomposite demonstrated a higher melting point (by 10 °C), greater thermal stability (by 10 °C), greater tensile strength (by 2 times), as well as three and two times reduction in oxygen and water vapour permeability, respectively, when compared to pure PHB [32].

Behera et al. [33] studied the effect of jute fibre with varying weight percentage (10–40 wt.%) on thermoplastic starch (TPS) by using compression moulding technique. Maximum tensile strength was displayed by the composite containing 30% jute (27.3 MPa). Measurements of contact angle and water absorption revealed that composites have a moderate hydrophobic characteristic [33]. Balla et al. [34] prepared the composites made up of thermoplastic starch (TPS) reinforced with chitosan nanoparticles using compression moulding method. Significant increases are shown in the tensile strength and Young's modulus, the water sorption of TPS-based composites is also significantly impacted by the addition of chitosan [34]. When it comes to nano-uses, bio-nano composites have subsequently emerged as prospective replacements for traditional nanoparticle (NP) applications in food packaging. Furthermore, the integration of these NPs gives good moisture barrier properties, which aids in the preservation of food quality throughout storage and transportation.

3 Biopolymeric Composites for Food Packaging Applications

The art, science, and technology of packaging enable the cost-effective delivery of commodities to final consumers. The advancement of food packaging technologies is essential for preserving food freshness. The food packaging industry has used biopolymer-based materials such as edible coatings/films, antioxidative and antibacterial films, and biodegradable films. The food quality and shelf life of food are both improved by the addition of various additives, such as nutrients, antioxidants, antibacterial agents, and colours, into these sustainable materials [35]. Figure 3 shows the different types of biopolymers used in the food packaging applications.

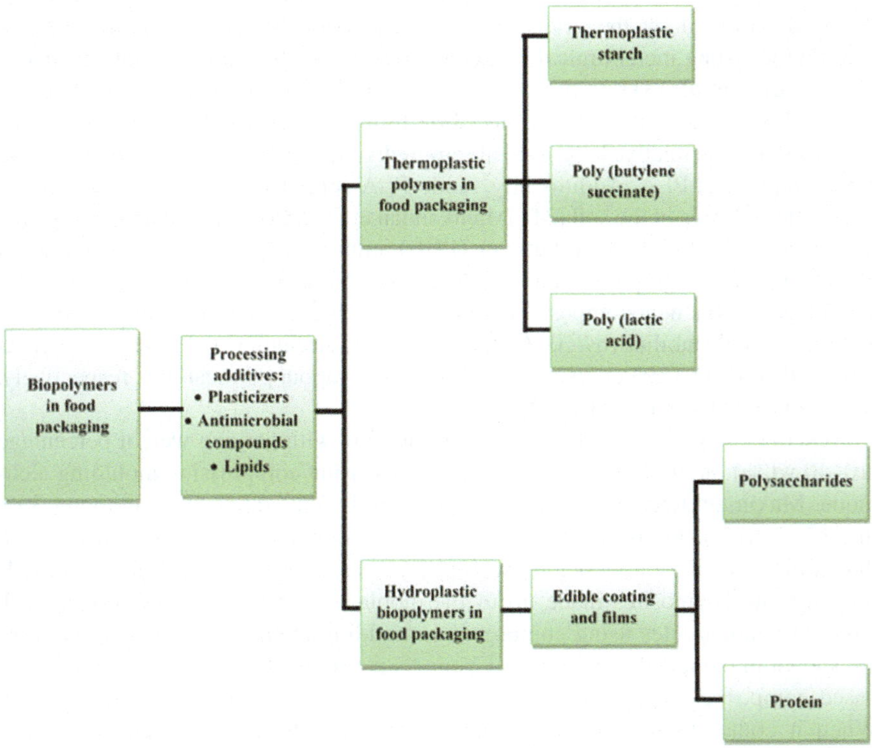

Fig. 3 Biopolymers classification in food packaging

3.1 Antimicrobial Packaging

Antimicrobial packaging is one type of method to restrict, decrease, impede, or kill the pathogenic micro-organisms in food. This approach has fascinated the food sector due to numerous benefits such as food safety, quality, and shelf life extension. Antimicrobial action is typically achieved by incorporating antimicrobial chemicals into the polymer matrix, which inhibits pathogen development. The most often used biodegradable polymers for active food packaging to overcome socio-environmental issues are chitosan, starch, and PBAT [36].

Chitosan-based biopolymers are used in packaging because of their biodegradability, antibacterial ability, biocompatibility, and ease of film formation. Roy et al. [37] reported the pullulan/chitosan-based multifunctional edible composite films were fabricated by reinforcing mushroom-mediated zinc oxide nanoparticles (ZnONPs) and propolis. The ZnONPs and propolis in the pullulan/chitosan-based biocomposite film provided significant antibacterial action against foodborne pathogens as well as excellent antioxidant activity. Pork belly packaging was done using the newly developed edible pullulan/chitosan-based film [37]. Zhang et al. [38] investigated the incorporation of TiO_2 nano-powder into chitosan in order to create a chitosan/TiO_2-based composite film. The findings revealed that chitosan/TiO_2 film wrapping on red grapes had effective antibacterial activity against various strains [38]. Ji et al. [39] studied the biodegradable chitosan-based film containing micro ramie fibre and lignin using casting method. The inclusion of various ratios of ramie fibre and lignin to the chitosan matrix resulted in a considerable improvement in mechanical, water resistance, thermal, and antioxidant properties. Overall, the chitosan-based films demonstrated significant food packaging potential [39].

Starch-based biopolymers have been widely used in the development of sustainable food packaging applications. To improve their water resistance, and mechanical and barrier properties, these polymers can be combined with additives. Tanwar et al. [40] developed an active antioxidant film with PVA and maize starch, as well as varying amounts of coconut shell extract (CSE) and sepiolite clay (SP). The oxidative stability of packaged soybean oil was increased by these PVA/starch film-based sachets [40]. Yusof et al. [41] prepared composite films based on polybutylene succinate (PBS), tapioca starch, and inclusion of Biomaster-silver (BM) nanoparticles. The composite film significantly improved the quality of chicken breast fillets by reducing colour variation, slowing pH rise, reducing weight loss, and delaying the hardening process [41]. Likewise, Romainor et al. [42] produced a starch/citrate film by using an antimicrobial agent called citric acid that was added in an optimised ratio. The findings showed that starch/citrate-based film effectively reduced 87–99% of fungal growth and 98–99% of foodborne bacteria development, extending the storage life of cake and bread products [42].

Active packaging applications require biopolymers to have antimicrobial properties, and these properties can be improved by adding various natural substances like organic acids, curcumin, pectin, nano-clays, and essential oils. Arumugam et al. [43]

developed the polybutylene adipate terephthalate (PBAT) biocomposite film incorporated with natural grape seed essential oil (GEO) in addition to silica nanoparticles (SiO_2 NPs) using the solution casting process. The inclusion of SiO_2NPs in the PBAT/GEO blend enhanced the tensile strength, thermal stability, and antibacterial activities whereas the incorporation of GEO enhanced the film's flexibility, opacity, and antimicrobial activity. Excellent antibacterial activity was shown by the PBAT/GEO/SiO_2NP films against food deterioration germs [43]. Montero et al. (2021) fabricated the PBAT active films loaded with cellulose nanofibers (CNF) embedded with cinnamon essential oil (EO) by the wire extension method. The findings established the physical interaction between the EO and the PBAT matrix, which modified the polymeric molecular configuration. The findings showed that the produced films enhanced the strawberry's characteristics and had antibacterial properties against Listeria monocytogenes and Salmonella, providing a viable replacement for synthetic packaging materials [44].

4 Market Trends of Biopolymers in Food Packaging

Innovative biopolymer formulations are being used to create unique sustainable materials that meet a variety of packaging needs, including food preservation and protection. Additionally, newly created biocomposites contributed to the improvement of barrier, antioxidant, mechanical, and antibacterial properties [45]. According to Unilever® survey in 2016, nearly 85% of customers are satisfied when purchasing food stuff in biodegradable bags. The biopolymer packaging industry is predicted to increase at a compound annual growth rate (CAGR) of 12.6% between 2021 and 2026 [46]. Consumers are especially concerned about the influence of the COVID-19 pandemic on food packaging, as the virus has the ability to remain on the plastic surface, so the need for antibacterial and antiviral property for food packaging material is essentially needed. Various commercially available products are successfully used in wraps made of sustainable biopolymers as shown in Fig. 4.

5 Future Scope and Conclusion

Biodegradable polymers made from natural biopolymers or synthetic bio-based polymers are gaining popularity in active food packaging applications, which minimises the consumption of plastic materials-based packaging.

- Food packaging is now responsible for the majority of environmental plastic waste. As a result, it is essential to allow the use of biodegradable packaging in commercial marketplaces in order to promote a sustainable and eco-friendly environment.

Fig. 4 Different biopolymer-based food packaging materials are available in markets

- Besides environmental problems, all food particularly fruits and vegetables has a reduced shelf life, when packaged in conventional plastic. Whereas, different edible films and coatings developed with various biopolymers effectively extended food shelf life.
- Nano-biocomposites extend food shelf life and improve the commercial viability of innovative packaging materials. These long-lasting packagings show outstanding antibacterial capabilities, particularly against *E. coli* and *S. aureus* bacteria, as well as high mechanical strengths and antioxidant activity.

Finally, it is suggested that the use of sustainable biopolymeric materials is crucial in managing plastic waste through the applications in food packaging sector.

References

1. Haghighi H, De Leo R, Bedin E, Pfeifer F, Siesler HW, Pulvirenti A (2019) Comparative analysis of blend and bilayer films based on chitosan and gelatin enriched with LAE (lauroyl arginate ethyl) with antimicrobial activity for food packaging applications. Food Packag Shelf Life 19:31–39

2. Silveira VAI, Marim BM, Hipólito A, Gonçalves MC, Mali S, Kobayashi RKT, Celligoi MAPC (2020) Characterization and antimicrobial properties of bioactive packaging films based on polylactic acid-sophorolipid for the control of foodborne pathogens. Food Packag Shelf Life 26:100591
3. Zhu, P., Wang, Y. and Zhang, X., 2022. Preparation and characterization of electrospun nanofibre membranes incorporated with an ethanol extract of Capparis spinosa L. as a potential packaging material. Food Packag Shelf Life 32:100851
4. Kabir E, Kaur R, Lee J, Kim KH, Kwon EE (2020) Prospects of biopolymer technology as an alternative option for non-degradable plastics and sustainable management of plastic wastes. J Clean Prod 258:120536
5. Chawla R, Sivakumar S, Kaur H (2021) Antimicrobial edible films in food packaging: current scenario and recent nanotechnological advancements-a review. Carbohydr Polym Technol Appl 2:100024
6. Hadidi M, Jafarzadeh S, Forough M, Garavand F, Alizadeh S, Salehabadi A, Khaneghah AM, Jafari SM (2022) Plant protein-based food packaging films; recent advances in fabrication, characterization, and applications. Trends Food Sci Technol
7. Hamed I, Jakobsen AN, Lerfall J (2022) Sustainable edible packaging systems based on active compounds from food processing byproducts: a review. Compr Rev Food Sci Food Saf 21(1):198–226
8. Chiralt A, Menzel C, Hernandez-García E, Collazo S, Gonzalez-Martinez C (2020) Use of by-products in edible coatings and biodegradable packaging materials for food preservation. In: Sustainability of the food system. Academic Press, pp 101–127
9. Atta OM, Manan S, Shahzad A, Ul-Islam M, Ullah MW, Yang G (2022) Biobased materials for active food packaging: a review. Food Hydrocoll 125:107419
10. Liu Y, Ahmed S, Sameen DE, Wang Y, Lu R, Dai J, Li S, Qin W (2021) A review of cellulose and its derivatives in biopolymer-based for food packaging application. Trends Food Sci Technol 112:532–546
11. Khalid MY, Arif ZU (2022) Novel biopolymer-based sustainable composites for food packaging applications: a narrative review. Food Packag Shelf Life 33:100892
12. Priyadarshi R, Roy S, Ghosh T, Biswas D, Rhim JW (2021) Antimicrobial nanofillers reinforced biopolymer composite films for active food packaging applications-a review. Sustain Mater Technol e00353
13. Basavegowda N, Baek KH (2021) Advances in functional biopolymer-based nanocomposites for active food packaging applications. Polymers 13(23):4198
14. Dharini V, Selvam SP, Jayaramudu J, Emmanuel RS (2022) Functional properties of clay nanofillers used in the biopolymer-based composite films for active food packaging applications-review. Appl Clay Sci 226:106555
15. Taherimehr M, YousefniaPasha H, Tabatabaeekoloor R, Pesaranhajiabbas E (2021) Trends and challenges of biopolymer-based nanocomposites in food packaging. Compr Rev Food Sci Food Saf 20(6):5321–5344
16. Kartik A, Akhil D, Lakshmi D, Gopinath KP, Arun J, Sivaramakrishnan R, Pugazhendhi A (2021) A critical review on production of biopolymers from algae biomass and their applications. Biores Technol 329:124868
17. Wang H, Ding F, Ma L, Zhang Y (2021) Edible films from chitosan-gelatin: physical properties and food packaging application. Food Biosci 40:100871
18. Karmanov AP, Kanarsky AV, Kocheva LS, Belyy VA, Semenov EI, Rachkova NG, Bogdanovich NI, Pokryshkin SA (2021) Chemical structure and polymer properties of wheat and cabbage lignins-valuable biopolymers for biomedical applications. Polymer 220:123571
19. Tang XZ, Kumar P, Alavi S, Sandeep KP (2012) Recent advances in biopolymers and biopolymer-based nanocomposites for food packaging materials. Crit Rev Food Sci Nutr 52(5):426–442
20. Basumatary IB, Mukherjee A, Katiyar V, Kumar S (2022) Biopolymer-based nanocomposite films and coatings: recent advances in shelf-life improvement of fruits and vegetables. Crit Rev Food Sci Nutr 62(7):1912–1935

21. Kumar P, Sandeep KP, Alavi S, Truong VD (2011) A review of experimental and modeling techniques to determine properties of biopolymer-based nanocomposites. J Food Sci 76(1):E2–E14
22. Kumar N, Kaur P, Bhatia S (2017) Advances in bio-nanocomposite materials for food packaging: a review. Nutr Food Sci 47(4):591–606
23. Rhim JW, Park HM, Ha CS (2013) Bio-nanocomposites for food packaging applications. Prog Polym Sci 38(10–11):1629–1652
24. Vieira MGA, da Silva MA, dos Santos LO, Beppu MM (2011) Natural-based plasticizers and biopolymer films: a review. Eur Polymer J 47(3):254–263
25. Rhim JW, Ng PK (2007) Natural biopolymer-based nanocomposite films for packaging applications. Crit Rev Food Sci Nutr 47(4):411–433
26. Li C, Wu J, Shi H, Xia Z, Sahoo JK, Yeo J, Kaplan DL (2022) Fiber-based biopolymer processing as a route toward sustainability. Adv Mater 34(1):2105196
27. Baranwal J, Barse B, Fais A, Delogu GL, Kumar A (2022) Biopolymer: a sustainable material for food and medical applications. Polymers 14(5):983
28. Aaliya B, Sunooj KV, Lackner M (2021) Biopolymer composites: a review. Int J Biobased Plast 3(1):40–84
29. Sobhan A, Muthukumarappan K, Wei L (2021) Biosensors and biopolymer-based nanocomposites for smart food packaging: challenges and opportunities. Food Packag Shelf Life 30:100745
30. Silvestre AJ, Freire CS, Vilela C (2021) Advanced biopolymer-based nanocomposites and hybrid materials. Materials 14(3):493
31. Sharma S, Majumdar A, Butola BS (2021) Tailoring the biodegradability of polylactic acid (PLA) based films and ramie-PLA green composites by using selective additives. Int J Biol Macromol 181:1092–1103
32. Manikandan NA, Pakshirajan K, Pugazhenthi G (2020) Preparation and characterization of environmentally safe and highly biodegradable microbial polyhydroxybutyrate (PHB) based graphene nanocomposites for potential food packaging applications. Int J Biol Macromol 154:866–877
33. Behera AK, Mohanty C, Pradhan SK, Das N (2021) Assessment of soil and fungal degradability of thermoplastic starch reinforced natural fiber composite. J Polym Environ 29(4):1031–1039
34. Balla B, Bartos A, Kun D, Csiszár E, Móczó J, Fekete E (2021) Improving mechanical and water sorption properties of thermoplastic starch by incorporating chitosan filler. Polym Testing 101:107278
35. Ilyas RA, Sapuan SM (2019) The preparation methods and processing of natural fibre biopolymer composites. Curr Org Synth 16(8):1068–1070
36. Thiyaku, T.T., 2022. Mechanical, barrier, thermal, surface hydrophobicity, optical and antibacterial properties of PBAT biopolymer after addition of natural plasticizer and silica nanoparticles. Pigment Resin Technol
37. Roy S, Priyadarshi R, Rhim JW (2021) Development of multifunctional pullulan/chitosan-based composite films reinforced with ZnO nanoparticles and propolis for meat packaging applications. Foods 10(11):2789
38. Zhang X, Xiao G, Wang Y, Zhao Y, Su H, Tan T (2017) Preparation of chitosanTiO_2 composite film with efficient antimicrobial activities under visible light for food packaging applications. Carbohydr Polym 169:101–107. https://doi.org/10.1016/J.CARBPOL.2017.03.073
39. Ji M, Li J, Li F, Wang X, Man J, Li J, Zhang C, Peng S (2022) A biodegradable chitosan-based composite film reinforced by ramie fibre and lignin for food packaging. Carbohydr Polym 281:119078
40. Tanwar R, Gupta V, Kumar P, Kumar A, Singh S, Gaikwad KK (2021) Development and characterization of PVA-starch incorporated with coconut shell extract and sepiolite clay as an antioxidant film for active food packaging applications. Int J Biol Macromol 185:451–461
41. Yusof NL, Mutalib NAA, Nazatul UK, Nadrah AH, Aziman N, Fouad H, Jawaid M, Ali A, Kian LK, Sain M (2021) Efficacy of biopolymer/starch based antimicrobial packaging for chicken breast fillets. Foods 10(10):2379

42. Romainor AN, Chin SF, Lihan S (2022) Antimicrobial starch-based film for food packaging application. Starch-Stärke 74(3–4):2100207
43. Arumugam S, Kandasamy J, Thiyaku T, Saxena P (2022) Effect of low concentration of SiO2 nanoparticles on grape seed essential oil/PBAT composite films for sustainable food packaging application. Sustainability 14(13):8073
44. Montero Y, Souza AG, Oliveira ÉR, dos Santos Rosa D (2021) Nanocellulose functionalized with cinnamon essential oil: a potential application in active biodegradable packaging for strawberry. Sustain Mater Technol 29:e00289
45. Mallakpour S, Azadi E, Hussain CM (2021) Recent breakthroughs of antibacterial and antiviral protective polymeric materials during COVID-19 pandemic and after pandemic: coating, packaging, and textile applications. Curr Opin Colloid Interface Sci 55:101480
46. Werner BG, Koontz JL, Goddard JM (2017) Hurdles to commercial translation of next generation active food packaging technologies. Curr Opin Food Sci 16:40–48

Improving the Sustainability of Autogenous Pulsed Current Gas Tungsten Arc Welding

M. Mohanram, Jayakrishna Kandasamy, G. Rajyalakshmi, and S. Arulvel

Abstract Pulsed current gas tungsten arc welding (PCGTAW) technology is one of the widely used welding techniques in many industries due to its high precision, excellent welding quality, and low equipment cost. However, PCGTAW is a parameter-dependent process and hence; there is a need for a prediction model to optimize the welding process parameters to enhance the performance and sustainability of the PCGTAW process. The present study evaluates the welding performance of SS409M with variation in depth of penetration using the design of experiments. Genetic algorithm (GA) and Simulated Annealing algorithms (SAA) were used to simulate, predict, and optimize the depth of penetration of the PCGTAW process. Peak current, base current, percent on time, pulse frequency, and welding speed were taken at three different levels. With the use of GA, the optimal welding parameters were found at a peak current of 171.802 A, pulse frequency of 4.748 Hz, 40% of the time, Base current of 65.691 A, and a welding speed of 60.001 mm/min. This combination has resulted in the depth of penetration of 4.12 mm, which is comparatively higher than the depth of penetration obtained for the SAA. Furthermore, a detailed study on the role of GA and SAA on the optimization of the depth of penetration in ensuring the sustainability of PCGTAW is reported in this study.

Keywords Optimization · Pulsed current gas tungsten arc welding (PCGTA) · Design of experiments (DoE) · Genetic algorithm (GA) · Simulated annealing algorithm (SAA)

M. Mohanram · J. Kandasamy (✉) · G. Rajyalakshmi · S. Arulvel
School of Mechanical Engineering, Vellore Institute of Technology (VIT), Vellore, Tamil Nadu 632014, India
e-mail: mail2jaikrish@gmail.com

M. Mohanram
e-mail: mohanram.m2021@vitstudent.ac.in

G. Rajyalakshmi
e-mail: rajyalakshmi@vit.ac.in

S. Arulvel
e-mail: arulvel.s@vit.ac.in

J. Kandasamy et al. (eds.), *Progress in Sustainable Manufacturing*, Management and Industrial Engineering, https://doi.org/10.1007/978-981-99-0201-9_9

1 Introduction

The growth of the manufacturing sector in any country is directly coupled with the GDP of that country. Due to stringent rules and environmental regulations, countries now focus on establishing robust and energy-efficient infrastructures in place for their effective utilization [1]. Energy-efficient measures and processes can bring about considerable cost savings too. Energy consumption in the manufacturing sector can be minimized by using contemporary energy-saving techniques. Globally, the manufacturing sectors consumed 2177 petajoules as part of their production activity in 2014, and over this same period, manufacturing sales increased by 15% [2]. Since 2019, despite the recovery in total industrial sales, energy consumption for all industries remains below that stated before the 2020 economic downturn, representing an enhancement in energy consumption. Sustainable consumption and production are also part of UN SDG goals (SDG-12) [2]. Sustainable energy consumption is the new norm of modern industries as part of their sustainability reporting activity. More than half of the products manufactured require welding. Aerospace, automotive, railroads, shipping, construction, and infrastructure, and other manufacturing industries heavily depend upon the welding process.

Welding is one of the significant methods in all sheet metal-oriented industries [3]. Gas Tungsten Arc Welding (GTAW) is used extensively for the production of unique components, including complicated structures like pressure vessels, offshore structures, nuclear reactors, petrochemicals, equipment for food processing, etc. In GTAW due to continuous heat input, the weld zones commonly show coarse grain structures. Also, the creation of Heat Affected Zone results in sub-standard welded joints, hot corrosion cracking, and results in loss of mechanical properties [3, 4]. PCGTAW process is one of the widely used welding techniques and proved to hold many advantages as compared to the GTAW. The benefits include regulating the total heat input, low distortion, reduced number of HAZ, residual stress, and micro-segregation [5]. In the PCGTAW process, the welding current is pulsated between low-and high-level periods. So, during main or pulse current, it melts down the weld region and allows to cool the molten pool and solidify all through the low or background current time [6]. The bead geometry is the number of overlapping weld spots and the number of overlaps is mainly determined by the pulse frequency [5].

Madhusudhan et al. [6] concentrated on optimization of the pulse frequency of PCGTA welds of Aluminium alloys for enhancing the mechanical properties by grain structure improvement. The improved UTS was reported to be attained under the 6 Hz pulse frequency of welded joints. Recently, many industries are adopting machine learning and evolutionary algorithms for manufacturing process optimization. Evolutionary approaches drive the elimination of unnecessary testing, save time, and reduce materials wastage and effort within the industries. Several computational models were introduced by many researchers to predict and optimize the manufacturing process responses. Mathematical models were developed to predict the process parameters based on the experimental data using statistical approach and ANN, etc. Based on the mathematical models developed, the adaption of evolutionary algorithms is very effective to optimize various manufacturing processes. Welding processes are no exclusion to the use of such algorithms to forecast and optimize the mechanical, geometrical, structural properties of the weldment before the welding process. Enhancement of weld quality and productivity is complicated since the optimization process in practice is extremely controlled by welding trials and depends on practical knowledge and involvement [7]. Welding business requires an effective means of parametric optimization due to the development of advanced metals and materials and increased demand for personalized design and applications.

The welding process is influenced by many process parameters, but to get the productive output, engineers adopt the Taguchi method for experimental design based on diversity in industrial applications. For process modeling, optimization, and parametric control, the adaption of the Taguchi method for the welding process has been proven to be very useful. Juang et al. [3] adopted the Taguchi method for optimizing the weld pool geometry of stainless steel using GTAW. But, the Taguchi method has some boundaries as it gives optimal solutions only within the bounded level of control factors. With continuous process parameters, it is difficult to predict the optimal values using the Taguchi method. For the prediction of optimal parameters, ANN is a preferable tool, and can precisely relate inputs with outputs [8, 9]. A trained ANN model can be used to calculate the responses accurately for specified parameter settings. Ciurana et al. [10] used the ANN approach for the prediction of preferable machining parameters to get the desired output in the pulsed laser micromachining process. In addition, Khaw et al. [11] explained the usage of the Taguchi concept for ANN design. For ANN design, first, it is the only known method that considers robustness as a significant design measure that helps in enhancing the quality of the ANN. Second, for the systematic design of the ANN model, the Taguchi method depends on orthogonal arrays (OAs) which drastically reduce the time for the design and development of ANN.

Many researchers have adopted evolutionary algorithms effectively for parametric optimization of various manufacturing processes. In every iteration, all the algorithms look for a locally optimal solution by stepping toward an improved solution. Hu et al. [12] adopted the PSO technique for many engineering applications to get the optimal solution. The original algorithm was slightly modified to retain the feasible solutions in the memory of the algorithm so that the set of solutions obtained can be reduced. It was determined that PSO is a capable approach for nonlinear optimization problems with inequality constraints. Katherasan et al. [13] also focused and succeed in the development of optimization models for a flux core arc welding process using PSO and ANNs to optimize the welding parameters. Roshan et al. [14] used the SAA for FSW process optimization to improve the mechanical properties of AA7075 welds. It was stated that the SAA algorithm is highly preferable for welding parameters optimization to obtain the desired output. Tarng et al. [15] used SAA to optimize the process parameters for the qualitative bead geometry. Further, the Fuzzy clustering technique was used for the classification of welds based on the quality of bead geometry. Many other researchers [16, 17] have also used SAA for optimizing the welding process to improve the quality of the weld bead geometry.

Similarly, to SAA and PSO, GA has also been extensively used for the optimization of parameters. Sathiya et al. [18] concentrated on optimization of the FSW process for enhancing the mechanical properties and simultaneously reduction of material loss during the welding process by adopting various optimization algorithms namely GA, SAA, and PSO. It was found that GA outperformed compared to all the other algorithms and the error percentage in results through GA compared to the experimental values was too less and negligible. Correia et al. [19] optimized the welding process for preferable deposition efficiency, bead width, depth of penetration, and reinforcement through GA and RSM. It was also stated that GA performs better than RSM with a proper setting of its internal parameters. Pashazadeh et al. [20] proved that GA is highly preferable for resistance spot welding (RSW) process optimization to predict the electrode tip dressing operation. All algorithms are suitable for welding process optimization, but they differ significantly with methodology, the computational approach, and the time required to achieve a viable solution.

Considering the importance of welding and the lack of optimization studies on PCGTA, this study aims at computing the optimal welding process parameters by which energy consumption can be reduced. Usage of computational methods such as GA and SAA forms the novelty of this study. This study also employed Design Expert in defining the objective function and the range of weld parameters used in GA and SAA.

2 Methodology and Experimentation

2.1 Material

SS409M is an alloy widely used across the automotive industries and also in other industrial applications [3]. The main applications of SS409M are in automobile mufflers, catalytic converters, tailpipes, farm equipment, structural supports, hangers, transformer cases, diamond tread plates, shipping containers, etc. It offers good corrosion resistance at high temperatures, good formability, and low cost. A 400 $\times$ 500 $\times$ 5 mm plate was sheared into 15 plates of each 100 $\times$ 70 $\times$ 5 mm dimension, such that each plate has enough size to accommodate three welding passes on it. The plates were filed to soften the edges and were then cleaned with emery sheets (with grit sizes 240/400/600 respectively) to get rid of rust for better weld quality. The plates were also cleaned with acetone. Figure 1. shows the workpieces after emery sheet polishing.

The chemical composition of SS409M is presented in Table 1 and the mechanical properties are listed in Table 2. The data required for the development of an optimization model to predict the Depth of Penetration (DoP) was obtained from welding of SS409M plates.

Fig. 1 Workpieces after emery sheet polishing

Table 1 Chemical composition of SS409M

Element	C	Mn	P	S	Ti	Cr	Ni
% composition	0.08	1.00	1.00	0.045	0.75	11.75	0.5

Table 2 Mechanical properties of SS409M

Property	Tensile strength (Mpa)	Yield Strength (MPa)	Elongation (% in 50 mm) min	Rockwell hardness (HRB) max	Brinell hardness (HB) max
Value	450	240	25	75	131

2.2 *Process Parameter Selection and Experimental Design*

The selection of process parameters for welding and their levels plays a key role in improving the quality of welding and improving process efficiency. To conserve energy in the welding process there exists a need to optimize the electrical parameters involved in it. In PCGTA welding the electrical parameters involved are peak current, base current, pulse frequency, and percent on time. Along with the electrical parameters, welding speed is also considered for experimentation. The set of process parameters and their levels are presented in Table 3.

Once the welding process, base material, and filler material were decided for experimentation, the next step is to plan for experimentation. The design of experiments approach is used to design the experimentation [22]. Orthogonality means that each parameter is evaluated independently and the effect of one factor does not

Table 3 Process parameters and their levels

Parameter	Level-1	Level-2	Level-3
Peak current (A)	160	170	180
Pulse frequency (Hz)	4	6	8
Percent on time (%)	40	50	60
Base current (A)	30	50	70
Welding speed (mm/min)	60	80	100

interfere with the estimation of the influence of another factor [22, 23]. Four factors (pulse current, background current, percent on time, pulse frequency, and welding speed) with three levels were selected as shown in Table 3.

The regression values of all parameters can be tested thoroughly only if a factorial design approach is used. For example, if there are '*n*' variables with each of one having '*m*' levels then the total number of combinations would be 'mn'. Factorial design is one of the simplest ways of approaching a problem where there are more than two variables [15]. In this case, there are five variables (*n*) with three levels (*m*) for each variable. The total combinations possible (mn) is 243. Handling all the 243 combinations and experimenting with them would be a tedious task, it also requires a lot of material (welding plates), consumes a lot of energy and time. A central composite design (CCD) was used for experimental design and the experimental plan was planned accordingly. By using the Response Surface Methodology (RSM) and adopting central composite design (CCD) the final 44 iterations were considered for the welding process and the results from the welding process were tabulated in Table 4.

The DoP obtained for each iteration was used as the input to the Design Expert (DE). (After processing the inputs, DE provided the optimal solution for DoE. The DS shows how each parameter will affect the welding process and it will also give the relation between any two parameters. The most influencing parameter on DoP can be predicted from the regression equation obtained from the DE. The objective function obtained from DE software was used in the GA and SAA to find the optimal welding parameters for welding SS409M using PCGTA.

3 Optimization of Experimental Results and Discussion

3.1 Prediction of Optimal Parameter Values

Based on the experimental results, the DoP is found to be maximum at the 38th iteration and the DoP value is observed to be low at the 32nd, 35th and 39th iterations. The values of the five factors influencing the DoP were given as input to the DE and the optimal parameter values for TIG welding/PCGTA welding of SS409M plates were obtained as tabulated in Table 5. The input parameters, influencing the DoP

Table 4 Design of welding experiment and results

Run	Peak current (A)	Pulse frequency (Hz)	Percent on time (%)	Base current(A)	Welding speed(mm/min)	DoP (mm)	Width (mm)
1	170	6	50	50	100	2.3	8.2
2	170	6	40	50	80	2.6	8.2
3	160	8	40	70	100	2.0	7.1
4	160	8	60	30	100	1.8	6.5
5	180	8	40	30	60	1.8	7.2
6	160	8	40	70	60	2.2	8.8
7	170	6	50	70	80	2.2	9.4
8	170	8	50	50	80	2.2	8.0
9	180	8	60	70	60	2.0	9.4
10	160	8	40	30	100	1.5	5.9
11	180	4	60	30	100	1.6	7.7
12	160	8	60	70	60	2.3	9.5
13	160	8	60	70	100	2.2	8.0
14	180	4	40	70	60	2.8	10.1
15	180	8	60	30	100	2.0	8.2
16	170	4	50	50	80	2.1	8.3
17	160	6	50	50	80	1.7	8.2
18	180	6	50	50	80	1.9	8.4
19	170	6	50	50	80	2.2	7.5
20	160	4	40	70	100	2.1	7.3
21	170	6	50	50	80	2.6	8.7
22	160	4	60	30	60	1.9	7.7
23	180	4	40	70	100	2.1	8.9
24	180	8	60	30	60	2.4	8.8
25	160	4	60	30	100	1.7	7.4
26	180	8	40	70	60	2.8	10.5
27	180	8	40	70	100	2.2	8.6
28	180	8	60	70	100	2.3	9.2
29	180	4	60	30	60	2.3	8.8
30	170	6	50	30	80	1.8	6.8
31	180	4	60	70	100	2.3	9.0
32	160	4	40	30	60	1.5	6.5
33	160	4	60	70	60	2.4	9.4
34	180	4	40	30	100	1.6	6.2
35	180	8	40	30	100	1.5	6.7

(continued)

Table 4 (continued)

Run	Peak current (A)	Pulse frequency (Hz)	Percent on time (%)	Base current(A)	Welding speed(mm/min)	DoP (mm)	Width (mm)
36	180	4	40	30	60	2.1	7.8
37	160	4	40	70	60	3.1	9.4
38	170	6	50	50	60	3.4	9.2
39	160	4	40	30	100	1.5	6.8
40	180	4	60	70	60	2.7	10.7
41	160	8	40	30	60	1.9	7.0
42	170	6	60	50	80	2.1	8.8
43	160	4	60	70	100	2.1	8.3
44	160	8	60	30	60	2.1	6.0

Table 5 Optimum values of PCGTA welding of SS409M plates

Peak current(A)	Pulse frequency(Hz)	Percent on time (%)	Base current(A)	Welding speed (mm/min)	Depth of penetration (DOP) mm
170.87	4.63	40	65.53	60.01	3.36349*

were taken as a base for finding the optimal parameter values using the DE and the optimal results were found at 170.8 A of peak current, 4.63 Hz of frequency of the pulse, the base current is at 65.53, welding speed maintained at 60.01 mm/min, and DoP at these values was computed as 3.36349 mm.

3.2 *Impact of Welding Process Parameters on Depth of Penetration*

The impact of process parameters (A) peak current, (B) pulse frequency, (C) percent on time, (D) base current, (E) welding speed on DoP were plotted using DE represented in Fig. 2. As the peak current varies from 160 to 180 A Fig. 2b, the DoP increases up to 170 A and decreases from 170 till 180 A. The pulse frequency Fig. 2a varies from 4 to 8 Hz. The DoP starts increasing from 4 to 4.63 mm, and gradually decreases from 4.63 to 8.00 mm. The maximum DoP is obtained when the peak current is 170.87 A and pulse frequency is 4.63 Hz. As the peak current varies from 160 to 180 A, the DoP increases till 170 A and decreases from 170 till 180 A, and the percent of time varies from 40 to 60%. The DoP starts decreasing from 40% touches the lowest at around 50% and then increases from 50 to 60%. The maximum DoP is obtained when peak current is 170.87 A and percent on time is 40%.

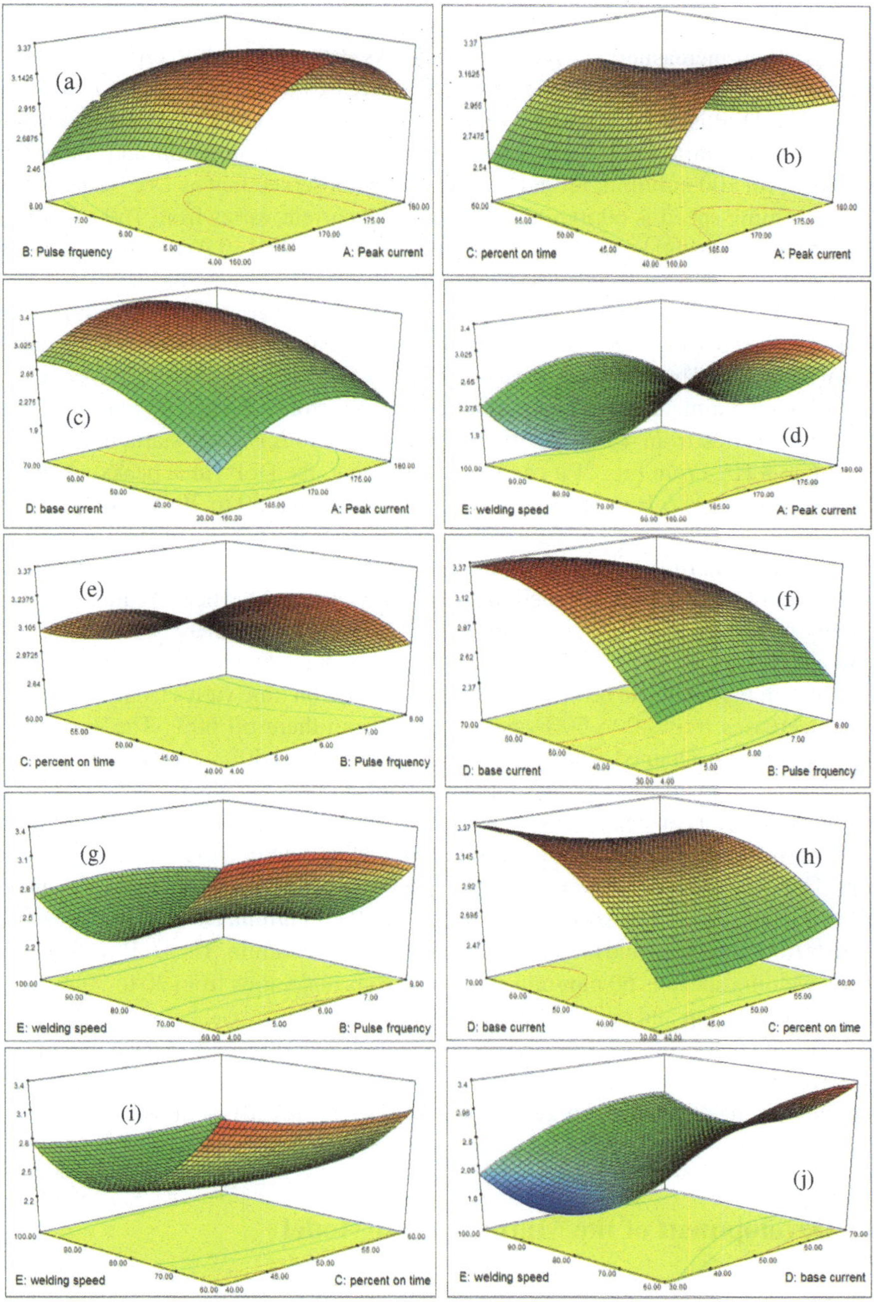

Fig. 2 Impact of process variables on the depth of penetration

From Fig. 2c it can be observed that, as the base current varies from 30 to 70 A, DoP goes on increasing and peaks at 70 A. As the peak current varies from 160 to 180 A, the DoP increases till 170 A and decreases from 170 till 180 A and the maximum DoP is at peak current values 170.87 A. As welding speed varies from 60 to 100 mm/min, DoP starts decreasing from 60 to 85 mm/min and thereafter increases till 100 mm/min. From Fig. 2d it can be observed that as DoP value peaks when welding speed is 60 mm/min. As the peak current varies from 160 to 180 A, the DoP increases up to 170 A and decreases from 170 till 180 A and the maximum DoP is at peak current values 170.87 A.

From Fig. 2e it can be noted that the percent of time varies from 40 to 60%, the DoP starts decreasing from 40% touches the lowest at around 50%, and then increases from 50 to 60%. The pulse frequency varies from 4 to 8 Hz, the DoP starts increasing from 4 to 4.63 mm, and results in a gradual decrease from 4.63 to 8.00 mm. As pulse frequency varies from 4 to 8 Hz, the DoP increases to 6 Hz and starts to decrease from 6 to 8 Hz. From Fig. 2f it can be observed that, the DoP value peaks when the pulse frequency is 6 Hz. As the base current varies from 30 to 70 A, DoP goes on increasing and peaks at 70 A. As welding speed varies from 60 to 100 mm/min, DoP starts decreasing from 60 to 85 mm/min and thereafter increases to 100 mm/min.

From Fig. 2g it can be detected that the DoP value peaks when welding speed is 60 mm/min. As pulse frequency varies from 4 to 8 Hz, DoP increases to 6 Hz and starts decreasing from 6 to 8 Hz. The DoP value peaks when the pulse frequency is 6 Hz. From Fig. 2h it can be observed that, as percent on time varies from 40 to 60%, DoP decreases from 40 to 50% and increases from there till 60%. The maximum value of DoP is obtained when the percent of the time is 50%. As the base current varies from 30 to 70 A, DoP goes on increasing and peaks at 70 A.

From Fig. 2i it can be noted that, as welding speed varies from 60 to 100 mm/min, DoP starts decreasing from 60 to 85 mm/min and thereafter increases till 100 mm/min. The DoP value peaks when welding speed is 60 mm/min. From Fig. 2j it can be observed that, as welding speed varies from 60 to 100 mm/min, DoP starts decreasing from 60 to 85 mm/min and thereafter increases till 100 mm/min. The DoP value peaks when welding speed is 60 mm/min. As the base current varies from 30 to 70 A, DoP goes on increasing and peaks at 70 A.

From the optimized set, Fig. 3 gives a clear understanding of how each parameter varies with DoP. From Fig. 3 it can be observed except welding speed (E) and percent on time (C) the other three parameters (A, B, D) vary directly with DoP.

4 Development of the Mathematical Model

The response function for DoP of the weld joints is a function of the peak current, pulse frequency, base current, percent of the time, and welding speed and it can be expressed as:

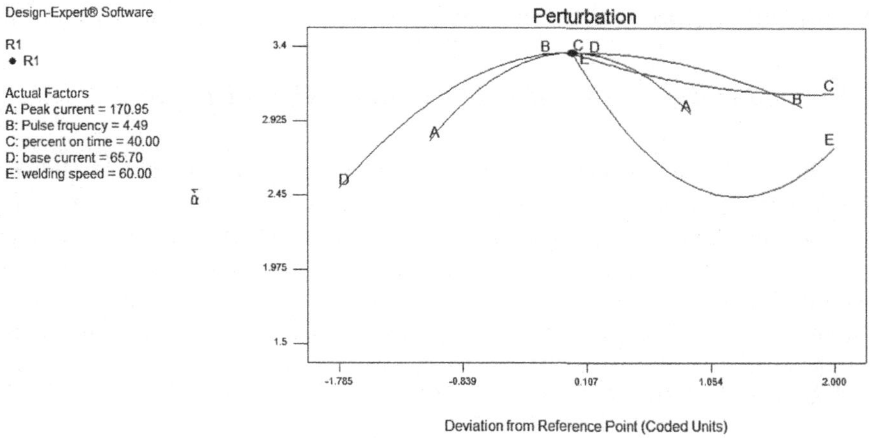

Fig. 3 The optimal set of process parameters for depth of penetration

$$DoP = f(\text{peak current, pulse frequency, base current, percent of time and welding speed}) \tag{1}$$

The process was mathematically modeled using regression analysis and the fit between DoP and input variables were represented (based on Eq. 1) as;

$$\begin{aligned}
\text{Depth of penetration} = {} & -129.58 + 1.63 * (\text{peak current}) \\
& +0.313 * (\text{pulse frequency}) - 0.077 * (\text{percent on time}) \\
& +0.13 * (\text{base current}) - 0.24 * (\text{welding speed}) \\
& -0.00031 * (\text{peak current} * \text{pulse frequency}) \\
& +0 * (\text{peak current percent on time}) \\
& -0.00009 * (\text{peak current} * \text{base current}) \\
& -0.00012 * (\text{peak current} * \text{welding speed}) \\
& +0.0016 * (\text{pulse frequency} * \text{percent on time}) \\
& -0.0018 * (\text{pulse frequency} * \text{base current}) \\
& +0.0014 + (\text{pulse frequency} * \text{welding speed}) \\
& -0.0053 * (\text{percent on time} * \text{base current}) \\
& +0.00025 * (\text{percent on time} * \text{welding speed}) \\
& -0.000016 * (\text{base current*welding speed}) \\
& -0.0047 * \left(\text{peak current}^2\right) - 0.03 * \left(\text{pulse frequency}^2\right) \\
& +0.00077 * \left(\text{percent on time}^2\right) - 0.0068 * \left(\text{base current}^2\right) \\
& +0.0014 * \left(\text{welding speed}^2\right)
\end{aligned} \tag{2}$$

5 Optimization of the Welding Process

5.1 Application of Genetic Algorithm for Optimization

With the inspiration of genetics and nature selection principles, in 1975, John Holland developed the GA for prediction purposes. The genetic algorithm starts with deciding the initial population. The population size of chromosomes can have a major influence on the outcome and also affects the time and computational work to get the optimized solution. After attaining the initial population, the error function for every chromosome can be calculated and the priority for the chromosomes can also be assigned based on the fitness and the error.

The steps involved in GA are (as adopted from):

Step 1. Generation of an initial population *M(0)* randomly
Step 2. Calculate and save the fitness *u(m)* for every individual m in the current population *M(t)*
Step 3. Define the range of probabilities *p(m)* for each *m* in *M(t)* so that *p(m)* is proportional to *u(m)*
Step 4. By selecting the individual probabilistically from *M(t), generate M(t + 1)* to produce offspring via genetic operators
Step 5. Step 2 is needed to repeat until an adequate solution is obtained.

The aim of using GA in this optimization process is to find the optimal values of process parameters that contribute to the maximum value of DoP. Optimization problem comprises of a minimization function defined by one of the second-order equation(s). The most important parameters to be considered are population size, the type of selection function, the crossover rate, and the mutation rate. By the process of trial and error, the value of the parameter can be computed for obtaining the most optimal result that is expected from this study.

5.2 Optimization of Depth of Penetration

The objective function is required for any optimization process, in this study for maximization objective of the optimization problem can be expressed as;

$$\text{Maximise DoP} = f\begin{pmatrix}\text{peak current, pulse frequency,} \\ \text{base current, percent on time and welding speed}\end{pmatrix}$$

Within the range of the welding parameters:

$$
\begin{aligned}
&160 < \text{peak current} < 180 \\
&4 < \text{pulse frequency} < 8 \\
&40 < \text{percent on time} < 60 \\
&30 < \text{base current} < 70 \\
&60 < \text{welding speed} < 100
\end{aligned} \tag{3}
$$

5.3 *Fitness Function for Depth of Penetration*

The fitness function of depth of penetration was developed based on mathematical model and represented as Eq. (4)

$$\text{function} y = \text{simple fitness}(x) \tag{4}$$

$$
\begin{aligned}
y =& (-1) * (-129.58 + 1.63 * x(1) + 0.31 * x(2) - 0.077 * x(3) \\
&+0.136 * x(4) - 0.24 * x(5) - 0.00031 * (x(1) * x(2)) \\
&+0.0 * (x(1) * x(3)) - 0.000093 * (x(1) * x(4)) \\
&-0.000125 * (x(1) * x(5)) + 0.00156 * (x(2) * x(3)) \\
&-0.0018 * (x(2) * x(4)) + 0.0014 * (x(2) * x(5)) \\
&-0.00053 * (x(3) * x(4)) + 0.00025 * (x(3) * x(5)) \\
&-0.0000156 * (x(4) * x(5)) - 0.0047 * (x(1)^2) \\
&-0.03 * (x(2)^2) + 0.00077 * (x(3)^2) \\
&-0.00068 * (x(4)^2) + 0.00144 * (x(5)^2))
\end{aligned} \tag{5}
$$

where, x(1) denotes peak current; *x(2)* denotes pulse frequency; *x(3)* denotes percent on time; *x(4)* denotes base current; and *x(5)* denotes welding speed

All objective functions, constraints, boundaries conditions, problem and function types were fed as input into the GA MATLAB code developed especially for this study. The optimized values obtained after all the iterations are presented in Table 6.

Within the range of the welding parameters (Eq. 3), GA will check for every combination possible. The regression equation obtained by welding the plates helped predict the DoP value for an unprepared iteration, tools like GA will help to overcome

Table 6 Optimized results obtained through GA

Peak current (amps)	Pulse frequency (Hz)	Percent on time (%)	Base current(A)	Welding speed (mm/min)	DoP (mm)
171.802	4.748	40	65.691	60.001	4.152

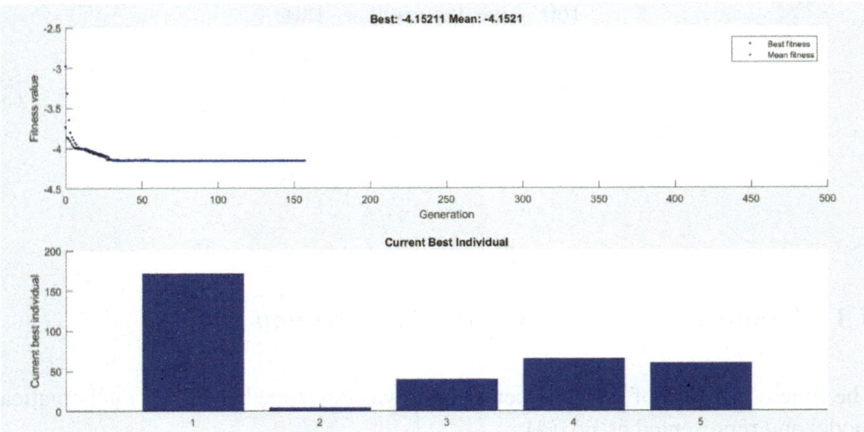

Fig. 4 Representation of optimal and best fitness value

this problem. GA MATLAB code performed around 157 iterations, Fig. 4 shows the number of iterations performed and the optimized values obtained for each parameter.

From Fig. 4, the maximum DoP from the GA analysis was found to be 4.152 mm at peak current 171.802 A, pulse frequency 4.748 Hz, percent on time 40%, base current to be 65.691 A, and welding speed 60.001 mm/min. But, as the effect of peak current is observed to be minimum, the process variable values generated by using GA can be considered to be optimal. From the percentage error for DoP, the experimental value is 3.3649 mm and the value generated by GA is 4.152 mm which shows a satisfactory improvement of 23.39% in the DoP.

5.4 Application of Simulated Annealing Algorithm for Optimization

In the SAA, only one solution is taken into account at one time. The fitness of the nearest solution is compared with the fitness of the current solution in SAA. Compared to the solution obtained, if the nearest solution fitness is better, then the nearest solution is accepted all the time. But, if the nearest solution fitness is poor than the current solution obtained, the nearest solution may be considered positively. The acceptance probability of acceptance of a poor solution depends upon the difference in error between the current solution and the two solutions [21].

In this study, the traditional SAA is used as an optimization tool to find the optimum values of the welding parameters to maximize the DoP. By using the coded values, the mathematical model was framed. A MATLAB code for the SAA was developed to optimize the welding parameters. Using the objective function and the range of welding parameters expressed by Eq. 3 and fitness function defined by Eq. 4 as inputs the optimized values were computed. With all objective functions, constraints, boundaries conditions, problem and function type fed as input into the SAA MATLAB code developed especially for this study, the optimized values obtained after all the iterations are presented in Table 7.

Within the range of the welding parameters (Eq. 3), SAA checks for each combination possible. Around 3700 iterations were performed using the SAA MATLAB code developed. Figure 5 shows how many iterations have been performed and the optimized values obtained for each parameter.

From Fig. 5, the maximum DoP based on the SAA was found as 4.048 mm at peak current 170.98amp, pulse frequency 5.384 Hz, Percent on time 40.154%, base current to be 63.844 A, and welding speed 61.204 mm/min. But, as the effect of peak current is observed to be minimum, the process variable values generated by using SAA can be considered to be optimal. From the percentage error for DoP, the experimental value is 3.3649 and the value generated by SAA is 4.048 which shows a satisfactory improvement of 20.30% in the DoP.

Table 7 Optimized values obtained from SAA

Peak current (amp)	Pulse frequency (Hz)	Percent on time (%)	Base current (amp)	Welding speed (mm/min)	DoP (mm)
170.981	5.384	40.154	63.844	61.204	4.048

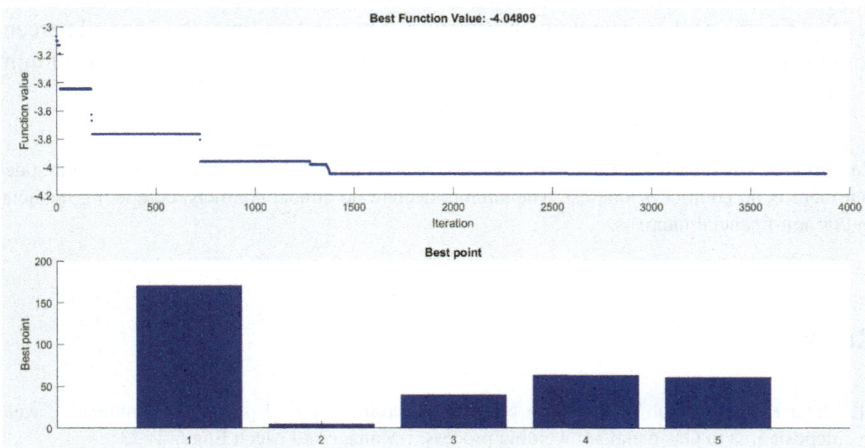

Fig. 5 Representation of optimal and best function value

6 Conclusions

The optimization of PCGTA welding using GA and SAA algorithms was explored in this work; more specifically, the prediction of the near-optimal PCGTA process parameters, peak current, pulse frequency, percent on time, base current, and welding speed. The optimum welding parameters were searched based on the minimization of an objective function, which considers all the cost-effective aspects (deposition efficiency) and the geometric characteristics (penetration, width, and reinforcement) of the bead. It was found that the GA can be a dominant tool in welding parametric optimization, even though there is no mathematical model available for the process. The key inferences of this study are:

- Peak current was found to be the most influencing parameter. The effect of parameters on DoP which are directly proportional are ranked in decreasing order as; Peak current > Pulse frequency > Base current.
- Welding speed and percent on time vary inversely to DoP. From the optimal results obtained, it can be inferred that there is not much change in the peak current value, so by setting peak current as 170 A maximum DoP can be achieved.
- While performing bead on plate welding, setting the base current as 50 A maximum DoP is achieved. But, according to GA when the base current is 65 A the maximum DoP obtained is 23.44% more than the maximum DoP obtained from experimental results and according to SAA when the base current is 63.84 A DoP obtained is 20.35% more.
- And finally, as there is not much variation in percent on time, the optimal percent on time value is 40%, the optimal welding speed can be maintained at 60 mm/min.

Compliance with Ethical Standards On behalf of all authors, the corresponding author states that there is no conflict of interest. The authors declare no ethical conflicts, competing financial and/or non-financial interests.

References

1. Babu KP, Chennaiah MB, Mohan NK (2016) Optimization of parameters influencing weld deposit area in Gas metal arc welding process. i-Manager's J Mech Eng 6(2):23
2. Lucas W, Howse D (1996) Activating flux-increasing the performance and productivity of the TIG and plasma processes
3. Juang SC, Tarng YS (2002) Process parameter selection for optimizing the weld pool geometry in the tungsten inert gas welding of stainless steel. J Mater Process Technol 122(1):33–37
4. Choudhury N, Bandyopadhyay A, Rudrapati R (2014) Design optimization of process parameters for TIG welding based on Taguchi method. Int J Curr Eng Technol 2:12–16
5. Harinadh V, Edison G, Akella S, Reddy LS, Buddu RK (2017) Multipass welding on inconel material with pulsed current gas tungsten arc welding. Mater Today: Proc 4(2):1452–1458
6. Reddy GM, Gokhale AA, Rao KP (1998) Optimisation of pulse frequency in pulsed current gas tungsten arc welding of aluminium–lithium alloy sheets. Mater Sci Technol 14(1):61–66

7. Bengtsson P (1993) Increasing productivity of metal-active-gas welding) it is possible. Praktiker 45(8):428–430
8. Su CT, Chiang TL (2003) Optimizing the IC wire bonding process using a neural networks/ genetic algorithms approach. J Intell Manuf 14(2):229–238
9. Kim IS, Jeong YJ, Lee CW, Yarlagadda PK (2003) Prediction of welding parameters for pipeline welding using an intelligent system. Int J Adv Manuf Technol 22(9–10):713–719
10. Ciurana J, Arias G, Ozel T (2009) Neural network modeling and particle swarm optimization (PSO) of process parameters in pulsed laser micromachining of hardened AISI H13 steel. Mater Manuf Process 24(3):358–368
11. Khaw JF, Lim BS, Lim LE (1995) Optimal design of neural networks using the Taguchi method. Neurocomputing 7(3):225–245
12. Blum C, Li X (2008) Swarm intelligence in optimization. In: Swarm intelligence. Springer, Berlin, Heidelberg, pp 43–85
13. Katherasan D, Elias JV, Sathiya P, Haq AN (2014) Simulation and parameter optimization of flux cored arc welding using artificial neural network and particle swarm optimization algorithm. J Intell Manuf 25(1):67–76
14. Roshan SB, Jooibari MB, Teimouri R, Asgharzadeh-Ahmadi G, Falahati-Naghibi M, Sohrabpoor H (2013) Optimization of friction stir welding process of AA7075 aluminum alloy to achieve desirable mechanical properties using ANFIS models and simulated annealing algorithm. Int J Adv Manuf Technol 69(5–8):1803–1818
15. Tarng YS, Tsai HL, Yeh SS (1999) Modeling, optimization and classification of weld quality in tungsten inert gas welding. Int J Mach Tools Manuf 39(9):1427–1438
16. Czyżak P, Jaszkiewicz A (1998) Pareto simulated annealing—A metaheuristic technique for multiple-objective combinatorial optimization. J Multi-Criteria Decis Anal 7(1):34–47
17. Kolahan F, Heidari M (2010) Modeling and optimization of MAG welding for gas pipelines using regression analysis and simulated annealing algorithm
18. Sathiya P, Aravindan S, Haq AN, Paneerselvam K (2009) Optimization of friction welding parameters using evolutionary computational techniques. J Mater Process Technol 209(5):2576–2584
19. Correia DS, Gonçalves CV, da Cunha Jr SS, Ferraresi VA (2005) Comparison between genetic algorithms and response surface methodology in GMAW welding optimization. J Mater Process Technol 160(1):70–76
20. Pashazadeh H, Gheisari Y, Hamedi M (2016) Statistical modeling and optimization of resistance spot welding process parameters using neural networks and multi-objective genetic algorithm. J Intell Manuf 27(3):549–559
21. McMullen PR, Frazier GV (2000) A simulated annealing approach to mixed-model sequencing with multiple objectives on a just-in-time line. IIE Trans 32(8):679–686
22. Yousefieh M, Shamanian M, Saatchi A (2012) Optimization of experimental conditions of the pulsed current GTAW parameters for mechanical properties of SDSS UNS S32760 welds based on the Taguchi design method. J Mater Eng Perform 21(9):1978–1988
23. Ross PJ, Ross PJ (1988) Taguchi techniques for quality engineering: loss function, orthogonal experiments, parameter and tolerance design (No. TS156 R12). McGraw-Hill, New York

An Integration of Smart Technology in Manufacturing

Raja Sreedharan and Tarik Saikouk

Abstract Nowadays, business practices are moving towards sustainability, which is necessary for organisational growth and environmental well-being. However, the world is becoming more VUCA making the business environment more volatile and uncertain. Moreover, complex decision-making and the human illusion of understanding of sustainability lead to ambiguous decision-making. Therefore, the present chapter explores the role of Industry 4.0 (IE4.0) and Big Data (BD) in Sustainable Manufacturing (SM) to understand the need for process visibility and enhance the decision-making for SM practices. To achieve this, the present chapter reviews scholarly work related to IE4.0, BD, and SM. From, the extant review unique constructs namely *SM in the VUCA world, Impact of Industry 4.0 in SM, and Big Data Architecture* were identified. The review will serve as reference material for practitioners and academicians in understanding the SM practices in the VUCA world.

Keywords Fourth industrial revolution · Big data · Sustainability · Manufacturing

1 Introduction

Nowadays, business practices are becoming more complex owing to the presence of different stakeholders and the increasing uncertainty in the operations, making the business practices more challenging to manage. Moreover, with the rise of information flow, customers' knowledge of business practices is increasing every day making the practices more volatile. As a result, the customer expects the company to provide products and services that are sustainable to the society. Further, to meet customer expectations, organisations have shifted towards sustainability practices

R. Sreedharan (✉)
Rabat Business School, UIR, Salé, Morocco
e-mail: rajasreedharanv@gmail.com

T. Saikouk
La Rochelle Business School, Excelia Group, CERIIM, 17000 La Rochelle, France
e-mail: saikoukt@excelia-group.com

J. Kandasamy et al. (eds.), *Progress in Sustainable Manufacturing*, Management and Industrial Engineering, https://doi.org/10.1007/978-981-99-0201-9_10

[1]. Furthermore, government policies and regulations has made sustainable business practices mandatory for business. However, most of the organisations are still uncertain as they are yet to explore the regulatory measures fully. In the context of sustainable business practices, the SM is considered as a frontrunner due to its global presence and its impact on our daily life. Moreover, organisations have started adopting emerging technologies like IE4.0, which enables the organisation to produce better products and services with minimum complexity. Also, there is a huge flow of information, business needs to embrace emerging technologies to avoid ambiguous decision-making. Such an environment represents a VUCA world and the SM is no exception to the VUCA world.

As business is facing a huge of information, which has led to a phenomenon known as BD. BD enables an organisation to make a proactive decision based on the data collected from different sources, leads to the effective management of data processing, and aids in creating a useful solution with far-reaching implications with minimum ambiguity. Moreover, IE4.0 focuses on the effective use of resources and work capacity, leading to increased production with the existing workforce. Such efficiency in IE4.0 and BD promotes the study to explore the impact of IE4.0 and BD [2] on the SM practices. Further, IE4.0 helps manage the supply chain in a better manner with facility insight and an integration method for inventory and data sharing through all levels, thus making decision-making efficient in a VUCA context. Therefore, an IE4.0 ecosystem can effectively drive sustainable practices in an organisation with better visibility and decision-making [3]. However, reporting on the impact of IE4.0 and BD on SM practices is found to be limited. In SM practices, various activities influence the decision-making process and having visibility over the process is instrumental for minimising process variation and effective decision-making [4, 5].

2 Review Findings

Therefore, the present review focuses on evaluating the impact of IE4.0 and BD in SM practices were reviewed in detail in the following sections using the content analysis.

2.1 SM in the VUCA World

Due to the rising uncertainty in the markets, organisations focus on controlling every step of their supply chain operations to meet the demand dynamics and ensure product delivery to the end customer. Further, to achieve better economic benefits, manufacturing and process industries are focused on reducing their carbon footprints significantly for environment well-being [6]. Moreover, most studies initially viewed environmental issues as a strategy problem. However, recent literature shows that the eco-innovation methods, such as life cycle valuations, clean technology, and eco-friendly design, are needed to be part of the organisations daily operations to minimise volatility in their process [7]. Therefore, industries started making their supply chain practices more eco-friendlier to the environment without compromising on the economic benefits [8]. Such a scenario was portrayed by organisations as SM. Later [4], described SM as "*the management of material, information and resource flow with the employee support in an organization along with the supply chain objectives on par with sustainability such as the environment, social and economic for the customer and shareholder benefits*". Further, sustainable development focuses on meeting current needs without comprising the future generations requirements. Therefore, SM focuses in operating in the VUCA world with the objective of being eco-friendly in the supply chain practices. Hence, it is necessary to explore the facets of SM and enhance them to develop an effective framework for decision-making. So, the facets of SM are reviewed in the following section.

2.2 Facets of SM

Economic: Economic studies on the supply chain deals with the specific solution to minimise the cost incurred in the supply chain operations. Most of the reviewed articles reported that the social and environmental objectives were secondary priorities, including the carbon emission, water recycling, and community jobs. Meanwhile, the strategy for eco-friendly practices to increase process visibility and decision-making is highly affected by the budget allocation [7]. Further, organisation needs to invest in technology to collect process information to make SM operations more robust and profitable to the organisation. Moreover, the studies show that decision-making related to the environmental and social dimensions of sustainability is challenging due to the nonlinear relationship among the dimensions which requires an evaluation approach [9].

Environment: In sustainable visibility, environment plays a major role with majority of the reviewed studies addressing issues related to the environment. The classifications of the environmental-based dimension are as follows: (1) material waste; (2) carbon footprints; (3) issues related to technological adoption; (4) greenhouse gas emission and the drive towards evaluating the dimension is challenging for many industries. As these dimensions are difficult to minimise and can increase to volatile level when influenced by uncertain events making their supply chain network more complex to operate with other stakeholders. Further, many studies reported on the life cycle assessment of the production systems and found that the organisation's transition to new production practices creates many ripples in their eco-friendly processes making it more volatile [10].

Social: To empower the local community, SM focuses on small businesses that depend on the supply chain operations for their survival. Further, studies show that the SM involves the local community, youths, and community members to create jobs in and around the supply chain hub, organisation, and processing centre. These practices will create a positive image for the organisation among the local community [11]. Employing locals and youths ensured strengthening the social aspect of SM. Moreover, choosing alternatives such as eco-friendly materials and training the community members can benefit the local community as it provides a positive image of the organisation. Therefore, such measures can promote the organisation more visibility and promotes the investor for new investment towards the organisation. Further, the governing body faces tough challenges in adopting alternatives. Also, new decision in SM influences the job opportunities for youth involvement, community well-being, training and development, and an assured stream of employment for the organisation [12]. So, there is need for an evaluation approach and framework for effective framework which can link the facets of SM in decision-making. Further, to understand the functional area of SM and its short-term and long-term impacts are presented in Table 1.

Table 1 Application of sustainability in the functional areas of SM

Application	Functional area	Description	References
Short term			
Reduce immediate negative impact on the environment	Environmental	Climate change and global warming are a result of the excess waste and emissions. Attributes of SM reduce this wastage and aim at creating an environment-friendly economy	[13]
Job creation	Economic	Local employment will go up, especially for the semi-skilled jobs	[14]
Animal health	Environmental	The overuse of resources adversely affects animal health	[3]
Compliance with the environmental law	Legal	Mandatory laws made by governments need to be followed, and SM helps organisations function within these laws	[4]
Long term			
Future revenue generation	Economic	Effective recycling and remanufacturing can increase the revenue generated eventually over a period	[7]
Improve product efficiency	Technological	Products within the attributes of SM will have increased life and, hence, higher value	[15]
Economic growth	Economic	Greater revenue generation and increased product life will contribute to the growth of the economy	[6]
Growing population	Societal	The population is exponentially increasing; this will put higher pressure on the existing resources	[16]

2.3 Impact of Industry 4.0 in SM

As the world is developing fast, industries have started evolving simultaneously with the emerging industrial practices, known as "Industry 4.0". The term "Industry 4.0" was proposed by a German association in 2011. This association includes practitioners, academicians, and legislators who put forth the concept of Industry 4.0. Industry 4.0 focused on improving the digital adoption intention in an organisation by adopting the best digital technologies [17]. Here, IE4.0 is driven by a wide range of technologies [2]. However, the main pillars of IE4.0 are as follows: (1) Autonomous System/Cyber-Physical Systems, (2) Simulation, (3) Horizontal and Vertical Integration, (4) Industrial Internet of Things (IIoT), (5) Cyber security, (6) Cloud Computing, (7) Additive Manufacturing, (8) Virtual Reality (VR), and (9) BD (10) Blockchain. As the market becomes competitive, organisations face tough challenges in IE4.0 adoption in their practices [18]. Section 3.2 discusses in detail BD and blockchain and how these can enhance the SM practices, while this section discusses the remaining pillars of IE4.0.

Cyber-physical system and simulation: Cyber-physical systems (CPS) involve an integration of computing, physical, and network processes. As the computer system gets embedded into the network, it starts monitoring the process in real time and gives feedback for process changes in the network. Such a system is known as CPS, and it has large potential for economic and societal applications in SM, and so organisations are investing heavily in CPS. Moreover, CPS combines the dynamics of physical processes and software system to provide precise analysis for better decision-making in the adopted network [3]. Apart from CPS, the simulation is widely used in SM, and it's very useful for process owner as it creates opportunity for the visibility of assumed conditions and enables decision-making for any given environment. Moreover, reviewed studies have used simulation to evaluate the condition in an SM. Most of the studies have reported three kinds of approaches for SM modelling: Discrete Event, System Dynamics, and Agent-Based Simulation. In these, the discrete event simulation is useful for the queuing system-based modelling and the agent-based system is used for modelling the action of the autonomous system (agent) in each network. Further, agent-based system can produce complex insights from the given network by collecting information on the agents' interaction, which are part of the network. Then system dynamics was used to model the complex nature of a given system. In system dynamics, the network is chosen, and this studies the network through stocks, flows, and feedback. These inputs are different from other forms, which explains the nonlinear relationship in a simple manner.

System integration and IIOT: In Industry 4.0, system integration is classified into two kinds: (1) Horizontal Integration; (2) Vertical Integration. From the review, it was noted that the majority of studies used the vertical integration and tried to create a better value chain from product manufacturing to sale distribution in SM. However, the horizontal integration was referred to in few works but has not been deployed yet [2]. The horizontal integration involves a better integration of end-to-end supply chain with the stakeholders and is built based on the organisation's core competencies and stakeholder's partnership. However, such an end-to-end value chain requires more investment and technological adoption, which have been reported in the reviewed works as agendas for future works.

Industrial Internet of Things (IIoT) is basically an extension of IoT for specific industrial applications. From the reviewed work, it is noted that IIoT, in an SM network is available for data collection and sharing. Then, it aids in analysis in the network and acts as a gatekeeper or tollgate and enables different devices to get connected to the SM network [19]. For deploying IIoT, the organisation should consider the following: (1) Focused deployment; (2) Point of usage; (3) Area of deployment. The IIoT usage has been extended to networks for different operations, such as monitoring the CPS, collecting the sensor information, reporting back to the network for real-time modifications (e.g., controlling the temperature in the furnace), optimising the fluid usage in machining, and scanning and grading of the raw material in the processing line. Further IIoT points of usage are subject to cost as the supply chain gets sophisticated with better sensor systems, but the cost incurred is high due to the machine's criticality and its usage in SM.

Cyber security and cloud computing: Very few articles have reported on the context of cyber security. However, to make the SM secure, cyber security can be used in the following: (1) Digital value chain network (2) Smart warehouse and factory. As the interconnected devices become common these days, cyber risks tend to evolve and change the way the supply chain operates [20]. Each organisation should adopt security protocols and standards in their network. As the breadth of cyber issues is increased, organisations must be secure in their operations with IT upgradation and provide platform for ethical hacking to identify bugs and vulnerabilities in the supply chain networks. Further, being resilient is important as it is essential to develop an exit strategy for the supply chain network if the system fails due to the cyber-attack. Cloud computing is a kind of service delivered as an application for data storage over the internet. These services are provided for personal and industrial applications in different forms. The data are stored in a physical location known as "data centre" and the hardware used to store the data is known as "cloud". Most of the cloud-based applications are pay-as-you-go basic ones, and supply chain networks have adopted them in most operations, which can eventually lead to a horizontal integration of the SM network. From the article reviews, it was found that the process visibility metrics have hardly been linked to cloud computing in their operations. This is because the stakeholders are resistant to adopting this strategy, as the cloud computing breaks the silos-only approach for data storage. Further, long-term investment in cloud computing could lead to cost saving and create more flexibility in SM practices.

Additive manufacturing and Virtual reality: Additive manufacturing focuses on producing parts and different components at an economic price for the consumers. Developing a design in a computer-based programme for designing (CAD), the output is generated to make different parts using a 3D printer. A 3D printer creates a part by printing every layer into a complete part. For 3D printing, materials such as metals, thermoplastics, ceramics, and other biochemical agents are used. Such manufacturing practices can help in developing prototypes and customised parts for the consumers in a short span of time. Further, the cost and time can be optimised for mass production as the flexibility in designing a new product is better in additive manufacturing. Such a drive enables the local supplier in the SM to procure parts at a lower price, thus leading to a community of localised manufacturing environments. Virtual Reality (VR) is one of the pillars of IE4.0 with vast potential for industrial applications [2]. However, in the review, it was found that it is sparingly used in the SM. VR is the bridge that connects the physical world with the virtual world. AR has started emerging in the field of supply chain for logistics activities where it is used to look at the chosen location and identify the order quantities. Other areas of VR are in the warehouse for knowledge on material handling information. Moreover, VR is used for maintenance in a smart factory by intuitively displaying manual information on site for diagnostics and troubleshooting. The present section critically reviews the IE4.0 pillars. However, businesses are becoming volatile in nature due to the VUCA world and IE4.0 pillars are needs to make less ambiguous decisions. Further, as the production system is evolving due to the customer needs and shorter product life cycle, huge investment for IE4.0 for organisation growth.

2.4 Need for Big Data Architecture in SM

As part of the current review, study has identified articles on information sharing for better visibility and decision-making for SM. This is because BD is considered part of IE4.0 and sparingly reported in SM. Therefore, study has reported various facets of BDA. Further, most of the BDA is developed with the blockchain technology.

Blockchain: Even though it is considered part of Industry 4.0, blockchain is widely used in the BDA environment for securing information and recording the transactions. As the name suggests, each block stores data and is interlinked with each other [21]. Thus, each block is distributed or decentralised in nature. Every transaction in blockchain consists of the following stages: (1) transaction must occur, (2) transaction must be verified; (3) transaction must be stored in block; (4) transaction must possess a hash. Once a transaction occurs, the user input is stored as a block for further evaluation. It also facilitates data segmentation for thousands of transactions that contain information on the shipment information, vendor preference, unit cost, logistics providers, and other critical information for a supply chain process. The second stage is to acknowledge the nature of the transaction and verify it. Further, in an organisation, the tailor-made BDA monitors every transaction and confirms its nature of entry, including the time, amount, unit value, currency exchange rate, stakeholders, suppliers, and nature of procurement to ensure transparency and credibility in the supply chain operations. The third stage is to store the information in blocks [15]. To do this, the data are verified with the standard entries, such as the digital signature, ID number, and Company code. Once verified and approved, the data are stored in blocks. Finally, each block has a unique value known as hash. Hash is a process of converting a given letter or number into the encrypted form of information, which is then accessed with the right access code (key for the encryption). Once a block is hashed, it can be stored in the blockchain, which serves as a digital ledger, monitoring different kinds of supply chain transactions that are encrypted and cannot be deleted, and these are also less prone to cyber-attacks due to the use of blocks and cryptographic algorithm.

Data source: Data sources ranges from social media to machine-generated data that are captured by the system. Most of the industrial operations attempt to adapt BD at an early stage. This will give the industries an upper hand in the production process and serve as data source for decision-making [22, 23]. Further, they can save a lot of energy with the utilisation of advanced technologies and can also customise on the production process [24]. The data content is usually structured, semi-structured, or unstructured. Data content gives them better information management of resources, such as inventory in the warehouse, supplier preference, product procurement, and unit cost for material. Such information categorisation and structuring are instrumental for managers to achieve SDVM in the SM.

Data store and format: Further, the data store focuses on the information storage in JSON, XML, and binary forms (PDF and word files) to enable retrieval as and when required. Also, other related files such as audio, video, image, and other formats are available [22]. Moreover, a data store helps in obtaining real-time data on the process and in visualising the work process in the supply chain. Moreover, the data policy in one organisation differs from that in another and is guided by the different technologies and systems. Also, the industrial structure and design for data storage and retrieval create better information management, thus leading to transparency among the stakeholders [25].

Data stage: The next step is data staging, which has three components: (1) data cleaning process; (2) data transformation; (3) data normalisations. The data are first collected and stored in the company repository. Using the stored data, cleaning is done to remove missing values and unidentified and incomplete data to turn it to a usable form. Then, the transformation ensures that the data are converted into an appropriate form for analysis. The last step is normalisation, where the data schema is checked for redundancy to decrease the computing time.

Data processing: The final step, in the BDA, is data processing, where the data are scrutinised either in batches or in real time. Many organisations have adopted the batch evaluation as the data are collected from different sources and across multiple points. Meanwhile, some systems analyse data in real time, where the computing platform allows a programmer to retrieve unbound stream of data to develop real-time application for effective decision-making in the SM [26].

The above discussion clearly shows that the BDA can enhance the data handling in SM practices and enhance the decision-making process which is secure and reliable with the help of blockchain. Such BDA can facilitate effective decision-making reducing the uncertainty in data quality and ambiguity in data processing making the decision more robust for SM.

3 Understanding the SM in the VUCA World

From the above discussion, SM is influenced by vast amount of the data from the suppliers and vendors. However, SM-based studies show that very few application frameworks have been reported to manage the data for effective decision-making. Even though, majority of the manufacturers use IE4.0 in their production system. It is critical to understand the role of each pillar and invest accordingly for cash strapped organisation. Furthermore, SM in VUCA world is affected by the human's illusion of understanding which leads to poor decisions. Therefore, enhanced SSM (Smart Sustainable Manufacturing) can facilitate SM to address the mentioned issues efficiently in the VUCA world. Following, this present section evaluates each structural attribute of SM and its role in developing an enhanced SSM application framework.

3.1 Linking the Facets of SM with IE4.0

Over the past few years, many works have articulated on SM practices in sufficient detail. However, SM is an extension of green supply chain management, which combines the ideas of the ecological philosophy, and the supply chain practices, including the constructs of sustainability, involving economic, social, and environmental aspects [19]. SM highlights an organisation's ability to use the material, energy, water, and other resources to discover solutions that can increase the performance of the supply chain process. Yet, in the VUCA world uncertainty and complexity of supply chain process makes it fuzzier for effective decision-making.

Even though, many initiatives have become digital in nature, with the adoption of Industry 4.0, kickstarting from the shop floor and converting the traditional factory into smart factory/connected smart factory (CSF). These digital technologies aid the organisation in generating links among their processes and provide systems the capabilities to complete the given process as per the client requirement. Further, it provides real-time functioning and marketplace information to the stakeholders. In the SM, digitalisation is achieved through the following features: connectivity, adoption, cognitive upgradation, integration, and autonomous control. These features can be achieved by the following digital technologies of Industry 4.0: Cyber-physical system (CPS), IIOT, Additive manufacturing, Augmented reality, System integration, Simulation, Cyber security, Cloud computing, BD, and Blockchain [27]. Also, it was proved that the IE4.0 offers several advantages, such as operation optimisation, value chain growth, and process control [28]. Therefore, the IE4.0 enables digitalisation and improves the SM operations both horizontally (functional parts) and vertically (across the entire value chain from procurement, production, processing to delivery to the customer [27].

However, majority of the works on SM focused on information flow and resource utilisation among the stakeholders, which are vital for organisational growth [3]. Furthermore, the economic aspect in SM is fulfilled by different stakeholders based on the supply chain operation, and the environmental aspect focuses on the green perspective of the SM practices. Finally, social aspect can enhance the societal attributes, which buttresses the local community thus leading to its empowerment (see Table 2). Further, SM practices are modified to complement sustainability through *dematerialisation* [29]. This refers to the lowest possible use of materials in production and cutbacks in material usage in design, packaging, or distribution in SM. The other significant manner to achieve this is through material optimisation and adopting emerging technology such as the IE4.0 and its pillars [5] which comprises of the following:

- Replacing expensive materials with economic one for data processing
- Using recyclable materials and stopping use of toxic materials using sensor system
- Curtailing scarce materials and promoting usage of eco-efficient materials
- Adopting of Industrial Internet of Things (IIoT) and sensor system to strengthen the SM practices.

Table 2 Linking the facets of SM with IE4.0

Attributes	IE4.0 pillars	Description	References
Design, manufacture, and distribute	Cyber-physical system	Involves usage of *dematerialisation*. The other significant manner is *optimisation* of materials. This has many facets	[29]
Usage—product life extension	IIOT	We extend the life of the product to its maximum. Increased life cycle will be the generator of profit rather than the point of sale. There are many methods to increase the life of the products	[30]
Recovery and recycling	System integration and BD	This attribute involves restoration of all that has been deemed as waste. It intends to cut down on waste or use waste to produce new materials	[31]
Remanufacturing	Additive manufacturing	Reconstitution and revival of the core products. Restoration to "Good as new" in full or in part. This can be done through many processes	[12]
Product as a service	System integration	There is a trade-off between ownership and cost that is used as an advantage. The model focuses on accessibility and not on ownership. It can be applied to a wide variety of products in many ways	[32]
Sharing platform	Blockchain	Promotes the idea of "co-ownership" or "co-access". Mainly aims at reducing the demand for newly manufactured goods	[33]

Further, Government norms and regulations ensure that the industries follow the guidelines and make their process more Eco-friendly [8]. In such, situation, overall process visibility is needed to monitor SM operations and avoid variation in following the guidelines [29]. For example, SM focuses on reducing the carbon footprint by transforming to digital platforms to save the consumable materials and it can monitor across the supply chain using the information sharing mechanism support by IE4.0 and enhance the visibility and decision-making to make actionable measures if a variation occurs in the stipulated guidelines for eco-friendly practices.

3.2 Utilising the Pillars of IE4.0 for Visibility and Decision-Making in SM

From the review, it was clear that the impact of IE4.0 and BD can lead to better visibility and decision-making for the SM practices. Further, the SSM can increase the SM efficiency, but different factors in the IE4.0 needs to be accounted for creating an ecosystem. The major factor is the denial of the transition towards digitalisation. While Firms think about transition, all traditional business models get affected and a new business model must be set up and framed for the smooth run and processing [34]. Firms will have to work with networking and build a complex network, and this will lead to many mergers and acquisitions with other firms, this is not an easy process and it is influenced by the firm size and business leaders will have to make as many strategic decisions to plan for the long run of the company and safeguard the firm's performance [14]. Moreover, to invest in IE4.0 technologies organisation must set up a guideline and allocate budget for deploying the pillar of IE4.0. Also, each pillar has its unique application, and its investment cost varies accordingly. From the review, study found that Cyber physical system (CPS); IIOT; Additive manufacturing are widely adopted in SM practices. Therefore, considered for the SSM evaluation. Meanwhile, organisation with more investment can focus on cloud computing and Blockchain for better digitalisation of SM practices.

However, literature evidence shows that IE4.0 adoption intention is still in the beginning stage and suppliers have started exploring the pillars of IE4.0. However, IT systems and communication models, which need to be linked to cloud space, which are prone to cyber threats during the organisation transformation towards IE4.0 [9]. Moreover, using IE4.0 for SSM involves lots of investment in terms of employee's skill making it a cumbersome process as it requires a specific training. Further, poor training programme and resistance to change will create uncertainty in the IE4.0 adoption and the stakeholder's involvement may drop, leading to a return to the traditional approach [26]. Moreover, developing the SSC's planning can be localised in terms of product procurement and resource allocation to reduce carbon footprint level. Further, data sharing in decision-making is highly influenced by the company policy, and there is a lack of trust among stakeholders, thus creating poor information visibility [8].

From, the above discussion, visibility in the SM process is instrumental to avoid the process becoming blurrier and more difficult to monitor for the organisation. Moreover, the different stakeholders at different level require better visibility on the SM practices for better decision-making and ensure process flexibility for SM operations [33]. Therefore, the SSM can ensure better visibility and decision-making along with flexibility for the SM operations using the pillars of IE4.0. Further, IE4.0 and BD complement each other; therefore, the study focuses on exploring the impact of IE4.0 and BD on SM practices [15]. Through extant literature review, it was found that BD is one of the most influential pillars of IE4.0 [23, 5]. Further, it generated more data for actionable insights for SM practices, making it more visible and effective for decision-making. For example, autonomous systems require sensors and a cyber-physical system to operate, and the generated data need to be stored in the cloud or Industrial Servers. Therefore, the IE4.0 driven system can rely on BD for effective decision-making in the VUCA world.

3.3 Enhancing Visibility and Decision-Making with Big Data Architecture

As the amount of data generated is increasing every day, BD refers to a mechanism of collecting and processing data, which is not considered feasible using the traditional methods. To handle BD, organisations used tailor-made designs for BD usage and known as BDA. However, BDA involves large data, which requires higher technological capability to collect, store, and process the information [15]. Further, BD can be expressed as the five Vs: Veracity, Velocity, Volume, Variety, and Value. As the *Volume* of data increases, the different insights such as pattern recognition, trends, and hidden information can be extracted with data analysis. A *Variety* of data is collected by the organisation from sensors, systems, IoT devices, and communication channels such as social media [35]. Such data range from images, audios, texts, and documents to logs, which are either structured or unstructured. Speed of data transfer influences the data content, and the *Velocity* of data is driven by constant absorption of data from different sources [23]. However, as data are collected from different sources, the biases and the noise factor in the data get increased with abnormality in data (*Veracity*). Therefore, the data need to be stored and retrieved meaningfully as per requirements. Further, veracity decides the value of the mined data for effective decision-making. Finally, the *Value* in BD refers to the process of finding huge hidden values from a given dataset with rapid response to business activities [27]. Therefore, BDA enables a business to collect data from multiple sources for effective decision-making. Further, the BDA is coupled with the blockchain technology to make the process more robust and secure from cyber threats and create accountability for the information storage, retrieving, and sharing from a specific repository. Therefore, coupling data coupled with blockchain can be vital for the BDA. Further, BDA can minimise the ambiguity in decision-making process. To achieve this, the following steps are needed:

Transparent system: In the supply chain activities, the information system is crucial for flawless operations and promotes the best business practices. However, the different stakeholders have systems that are not integrated with their data repository and have minimum access to the other players in the supply chain process. Therefore, a BD-based system, which is transparent and applicable to all the players in the supply chain process, needs to be constructed [3]. Moreover, it can link the data repository for effective decision-making for the stakeholders at different maturity levels, thus achieving enhanced SSM.

Experimentation: When BD is deployed, it can enhance the performance of both the internal and external customers through SSM. However, to develop an information system for SSM, organisations must do the required experimentation. As BD is being deployed, it creates the necessity for data growth and creation of information in a timely manner and this information needs to be vetted and approved for usage [17]. Moreover, experimentation enables the identification of vulnerability in the system and exposes them to help improve the system performance and information sharing across the platform.

Population segmentation: In the SM practices, segmenting the information as per the population is mandatory. Further, information on the environmental, social, and economic practices from different supplier locations can be marked and categorised for obtaining an actionable plan for product development [5]. Moreover, the procurement and government policies vary with the geographical locations, and these can be measured to foresee the pattern in the purchase, consumption, and utilisation of the product across the supply chain network.

Automated algorithm: Nowadays, the amount of data generated is of huge volume for each supply chain activity. However, it can be found, sorted, and segmented to obtain useful insights. To achieve this, organisations are using bot operations and automated algorithms. Most of the algorithms are tailor-made as per the field of application. It starts with data labelling, categorising the data points, and creating features from the data for further analysis [2]. In the following sections, the authors have explored the pillars of IE4.0 the SM practices.

So mentioned features of BDA can be designed in such way for SM operations to provide an upper hand in the production process and serve as data source for decision-making. Second, data store focuses on the information storage in JSON, XML, and binary forms (PDF and word files) to format the data and categorise their features to obtain information in real-time data leading to better transparency among the stakeholders. Then data stage ensures characteristics of stored data is structured for decision-making and minimising the ambiguity in the data through data normalisation. Finally, based on the nature of SM operations in each organisation processing the operational data to enhance the visibility of the value chain in SM and all these operations are backed by blockchain which monitors all transactions which creates authenticity and facilitates credibility for SM transactions in the VUCA world. Also, data generated from the value chain are staggering and require constant monitoring and evaluation. In such a scenario, BD can be used to label the information developed, which can then be vetted for actionable insights for SM operations making more certain and less ambiguous in decision-making.

4 Discussions and Findings

This work presents SM practices, which is now widespread in business development and policy debate on the sustainable growth of industrial production in the VUCA world. Moreover, policy development agencies and business associations view SM as an essential mechanism to endorse sustainable production, which will happen in the industrial transformations. The expectation is that the selection of SM will change economic activities away from dependence on non-renewable and emission-intensive carbon flows to more sustainable production and consumption. A similar study conducted by Liu et al. highlighted SM as a contestant for sustainable development through literature support. However, various attributes of SM, such as benefits of adoption of SM, success and failure factors, as well as SM applications, were not discussed. Moreover, this study extended the concept for BDA with a more straightforward framework for the application for sustainability visibility and decision-making. Later, Osterrieder et al. conducted a systematic review to address the spread of IE4.0 in sustainability. However, there was a need for the present works to extend the study further in the context of IE4.0 and BDA.

Therefore, the present focused on exploring the integration of Pillars of IE4.0 and BD for SM. From this, study can infer that the principal focus of the research has been on theoretical development, with very less attention to mathematical modelling-based works. Further, this was reemphasised by most theoretical papers obtained for the study. This is indicative that the concept is evolving into ideas concerning quantifiable numbers, measurements, and benchmarks. Further, the application areas studied show that maximum papers focus on "Industry 4.0" individually, directly contributing to the literature. Further, majority of the studies were simulation-based or literature reviews on the manufacturing front. However, very few were on the supply chain-related processes. Of the key attributes of SM, only two components refer directly to the organisation (social and environmental) with others focusing on services and sharing (economic).

From the review findings, the study developed unique structural attributes of SM practices namely: *SM in the VUCA world, Facets of SM, Pillars of Industry 4.0, and Big Data Architecture.* These attributes were reviewed through content analysis to understand the need for integrating IE4.0 and BDA for effective decision-making in SM practices. Also, the role of IE4.0 pillars in terms of investment point of review reviewed and found that Industries with minimum monetary support can focus on cloud computing and blockchain for data handling and sensor setup for physical attributes for developing an application framework. Moreover, in SM stakeholder plays a crucial role and it consists both the internal and external customers and stakeholder's maturity decides the nature of the SM operations. However, the stakeholder's involvement in the SM practices is influenced by various factors such as the organisational performance, employee support, training, knowledge of eco-friendly practices, and resource availability in the organisation. Therefore, present study linked the facets of SM with IE4.0 to facilitate effective decision-making to minimise process uncertainty. Also, Data collection is vital to get better visibility of the nature of process involved and stakeholder contribution towards the SM operations. Therefore, integration of IE4.0 and BD can create the opportunity to develop the data repository for the stakeholders related activities for effective decision-making in the SM practices.

4.1 Limitation and Scope for Future Research

The views of SM are vast and appealing, and the benefits of understanding the application framework at various levels are necessary for advancing SM practices among the stakeholders. At the industrial level, multidisciplinary methods linking business perspectives, policies, and technological developments need to be considered. Moreover, implementation of SM would begin with organisational development with due respect to change management and managerial mindset. Very few studies have developed the research design that orients to the need of the IE4.0 ecosystem for business and understands the effects of SM models for product, service, refurbishment, leasing, and remanufacturing. Therefore, researchers can focus on developing a tailor-made IE4.0 ecosystem with the right pillars of IE4.0 based on the resource availability and budget allocation.

The study has few limitations of its own. The study did not assess the link between the facets of SM with the Eco-Industrial Parks (EIP) and Industrial Symbiosis (IS). This would broaden the SM constructs for the SSM evaluation making it more robust. Besides, the SM framework needs to be further tested for generalisability in industrial settings. This would provide adequate information to policymakers for SM practices in the VUCA world. Additionally, the integration of IE4.0 and BDA demands employee adaption to IE4.0. However, the organisation must invest in their employees for better training and increase their loyalty to the organisation through job security. However, such initiatives have not been observed in the present study. Hence, researchers can conduct studies focusing on the employee training specific for the integration of IE4.0 and BDA.

In the SSM level and efficiency evaluation, the study has not addressed the non-linearity among the SM constructs which can influence the decision-making approach. Therefore, future researchers can develop an MCDM-based approach for capturing the multiple dimensions of SM can be used to explore the nature of relationship among the structural dimensions of the study. For the application framework, different agents, such as the hardware system, software, sensor setup, and programmable circuit, must procured based on the tailor-made operations. Therefore, future researchers can focus on extending the application framework for vendor selection for material procurement. Moreover, the organisation's inclination towards cyber security seems to be high. However, the literature reporting on this is found to be limited. So, further studies should focus on the nature of cyber security and its initiatives in the organisation, and how an organisation overcomes cyber-attacks can make for an interesting read.

5 Conclusion

The present work focused on the integration the IE4.0 and BD for enhanced SSM in SM practices. The work focused on understanding the need for effective decision-making and process visibility in the SM practices. The study critically reviewed the attributes of the IE4.0 and BD in the SM context. From, the review it was found that pillars of IE4.0 create better flexibility to the SM practices. Further, BDA ensures that seamless data processing for effective decision-making. Furthermore, the researchers can develop new application framework to achieve the SSM for industries. Along with this, the researchers developed an evaluating approach to measure the level and efficiency of SSM in an organisation. From, the application framework and the evaluation approach the visibility in SM can be significantly improved and the application framework can ensure better decision-making. Further, the benefits of SM are not only constrained to the environment but also offer plenty of economic benefits, such as employment, revenue, and future growth of the economy. As the SM practices and applications are dependent upon "governance" in a big way, the challenges, as well as hurdles, are dependent on the public governance challenges and community contribution that must be dealt with beforehand by being addressed adequately in time. Further, process transformation promotes cost savings and empowers the local communities and their well-being. However, achieving the process transformation for SSM is instrumental and it requires a case orgnisation to test the application model. Although study has taken adequate measures to avoid mistakes during the review, mistakes may have occurred in the reporting of the study.

References

1. Kumar S, Luthra S, Haleem A (2014) Critical success factors of customer involvement in greening the supply chain: an empirical study. Int J Logist Syst Manag 19:283–310
2. Rejikumar G, Arunprasad P, Persis J, Sreeraj KM (2019) Industry 4.0: key findings and analysis from the literature arena. Benchmarking 26(8):2514–2542
3. Chalmeta R, Santos-deLeón NJ (2020) Sustainable supply chain in the era of industry 4.0 and big data: a systematic analysis of literature and research. Sustainable 12(10):4108
4. Seuring S, Müller M (2008) From a literature review to a conceptual framework for sustainable supply chain management. J Clean Prod 16(15):1699–1710
5. Yadav G, Luthra S, Jakhar SK, Mangla SK, Rai DP (2020) A framework to overcome sustainable supply chain challenges through solution measures of industry 4.0 and circular economy: an automotive case. J Clean Prod 254:120112

6. Lezoche M, Panetto H, Kacprzyk J, Hernandez JE, Alemany Díaz MME (2020) Agri-food 4.0: a survey of the supply chains and technologies for the future agriculture. Comput Ind 117:103187
7. Munodawafa RT, Johl SK (2019) Big data analytics capabilities and eco-innovation: a study of energy companies. Sustain 11(15):4254
8. Craig T, Polhill G, Colley K, Carrus G, Maricchiolo F, Bonaiuto M, Bonnes M, Dumitru A, Mira RG (2019) Transmission of pro-environmental norms in large organizations. Sustain Prod Consum 19:25–32
9. Papetti A, Marconi M, Rossi M, Germani M (2019) Web-based platform for eco-sustainable supply chain management. Sustain Prod Consum 17:215–228
10. Seuring S, Gold S (2012) Conducting content-analysis based literature reviews in supply chain management. Supply Chain Manag 17(5):544–555
11. Tsai WH, Lu YH (2018) A framework of production planning and control with carbon tax under industry 4.0. Sustainable 10(9):3221
12. Andersson FNG (2020) Effects on the manufacturing, utility and construction industries of decarbonization of the energy-intensive and natural resource-based industries. Sustain Prod Consum 21. https://doi.org/10.1016/j.spc.2019.10.003
13. Luthra S, Mangla SK, Xu L, Diabat A (2016) Using AHP to evaluate barriers in adopting sustainable consumption and production initiatives in a supply chain. Int J Prod Econ 181:342–349
14. Sindhu SP, Nehra V, Luthra S (2016) Recognition and prioritization of challenges in growth of solar energy using analytical hierarchy process: Indian outlook. Energy 100:332–348
15. Sanghavi D, Parikh S, Raj SA (2019) Industry 4.0: tools and implementation. Manag Prod Eng Rev 10:3–13
16. Luthra S, Kumar S, Garg D, Haleem A (2015) Barriers to renewable/sustainable energy technologies adoption: Indian perspective. Renew Sustain Energy Rev 41:762–776
17. Gawankar SA, Gunasekaran A, Kamble S (2020) A study on investments in the big data-driven supply chain, performance measures and organisational performance in Indian retail 4.0 context. Int J Prod Res 58(5):1574–1593
18. Rajput S, Singh SP (2019) Identifying industry 4.0 IoT enablers by integrated PCA-ISM-DEMATEL approach. Manag Decis 57:1784–1817
19. García-Herrero L, De Menna F, Vittuari M (2019) Sustainability concerns and practices in the chocolate life cycle: integrating consumers' perceptions and experts' knowledge. Sustain Prod Consum 20:117–127
20. Rossit DA, Tohmé F, Frutos M (2019) A data-driven scheduling approach to smart manufacturing. J Ind Inf Integr 15:69–79
21. Manimuthu A, Raja Sreedharan V, Rejikumar G, Marwaha D (2019) A literature review on bitcoin: transformation of crypto currency into a global phenomenon. IEEE Eng Manag Rev 47(1):28–35
22. Afsari K, Eastman CM, Castro-Lacouture D (2017) JavaScript Object Notation (JSON) data serialization for IFC schema in web-based BIM data exchange. Autom Constr 77:24–51
23. Bibri SE (2018) The IoT for smart sustainable cities of the future: an analytical framework for sensor-based big data applications for environmental sustainability. Sustain Cities Soc 38:230–253
24. Hofmann E, Brunner JH, Holschbach E (2019) Research in business service purchasing: current status and directions for the future. Manag Rev Q 70(3):421–460
25. Fareri S, Fantoni G, Chiarello F, Coli E, Binda A (2020) Estimating Industry 4.0 impact on job profiles and skills using text mining. Comput Ind 118:103222
26. Ing TS, Lee TC, Chan SW, Alipal J, Hamid NA (2019) An overview of the rising challenges in implementing industry 4.0. Int J Supply Chain Manag 8:1181–1188
27. Ardito L, Petruzzelli AM, Panniello U, Garavelli AC (2019) Towards industry 4.0: mapping digital technologies for supply chain management-marketing integration. Bus Process Manag J 25:323–346

28. Strange R, Zucchella A (2017) Industry 4.0, global value chains and international business. Multinatl Bus Rev 25(3):174–184
29. Lieder M, Rashid A (2016) Towards circular economy implementation: a comprehensive review in context of manufacturing industry. J Clean Prod 115:36–51
30. Kausar K, Garg D, Luthra S (2017) Key enablers to implement sustainable manufacturing practices: an Indian insight. Uncertain Supply Chain Manag 5(2):89–104
31. Mangla SK, Luthra S, Jakhar S, Gandhi S, Muduli K, Kumar A (2020) A step to clean energy-sustainability in energy system management in an emerging economy context. J Clean Prod 242:118462
32. Karlsson NPE (2019) Business models and business cases for financial sustainability: insights on corporate sustainability in the Swedish farm-based biogas industry. Sustain Prod Consum 18:115–129
33. Torkabadi AM, Pourjavad E, Mayorga RV (2018) An integrated fuzzy MCDM approach to improve sustainable consumption and production trends in supply chain. Sustain Prod Consum 16:99–109
34. Haleem A, Mannan B, Luthra S, Kumar S, Khurana S (2019) Technology forecasting (TF) and technology assessment (TA) methodologies: a conceptual review. Benchmarking 26(1):48–72
35. Luthra S, Mangla SK, Yadav G (2019) An analysis of causal relationships among challenges impeding redistributed manufacturing in emerging economies. J Clean Prod 225:949–962

Embracing New Digital Technologies to Ensure Sustainability in Manufacturing

Soumya Prakash Patra, Sungjemmenla, and Rohit Agrawal

Abstract In the global emerging scenario, industries over wide sectors are adopting digital technologies to sustain in the market and to get a competitive edge over competitors. Nowadays companies are focusing on enhancing their technologies to achieve sustainable benefits. Digital technologies are the prime focus of today's industrial sector. The goal to achieve sustainable benefits can be effectively achieved by adopting several available digital technologies such as artificial intelligence, big data, cloud computing, and additive manufacturing. This study discusses various available digital technologies for their applications in the manufacturing sector to achieve sustainable benefits. Eight different digital technologies have been considered and discussed in this study. This study will enable industrial practitioners to identify the specific areas in manufacturing where digital technologies can be adopted to enhance business performance.

Keywords Digital technologies · Sustainable manufacturing · Industry 4.0 · Smart manufacturing

1 Introduction

The current globalization is confronted with sustainability issues from all three perspectives, including the economic, environmental, and social facets [27]. Sadly, there are more than one billion people living in extreme poverty, and income disparity is also rising. The main causes of the problem are unsustainable production and

S. P. Patra
General Management, Indian Institute of Management, Bodhgaya, Bihar 824234, India

Sungjemmenla
Department of Energy Science and Engineering, Indian Institute of Technology Delhi, New Delhi 110016, India

R. Agrawal (✉)
Operations Management and Quantitative Techniques, Indian Institute of Management, Bodhgaya, Bihar 824234, India
e-mail: mailerrohit@gmail.com

J. Kandasamy et al. (eds.), *Progress in Sustainable Manufacturing*, Management and Industrial Engineering, https://doi.org/10.1007/978-981-99-0201-9_11

consumption, which are endangering human life on Earth and causing serious socioeconomic difficulties. By 2050, the global population is anticipated to reach 9.5 billion people [1].

As a result of increasing manufacturing due to rising consumer demand, there is more output and business overall. This system has the potential to seriously harm the environment due to the use of non-renewable inputs, their increased rate of consumption, which is not sustainable, as well as higher emissions from the production process and the improper disposal of end-of-life items. Only by following the road of sustainable development, these issues can be solved. Thus, altering consumption and production habits can aid in protecting the environment and its natural resources. Manufacturers must adopt a new paradigm known as the circular economy (CE) in order to accomplish sustainable development goals [7].

However, a number of obstacles are preventing CE from being implemented and developing. The primary difficulties include expedited initial cost of setup, complexity in the supply chain, less cooperation between businesses, lack of information in designing a product and its manufacturing process, lack of skills, quality compromises, lengthy lead times for disassembly, and high costs associated with such processes [15]. Adopting digital technologies will help solve these problems.

A growing degree of digitalization in recent years has altered industrial production systems, resulting in smart, networked, and decentralized manufacturing. This new organizational level is frequently referred to as "The fourth industrial revolution" or "Industry 4.0" [21]. Industry 4.0's central premise is to leverage new digital technologies so that industrial practices and business processes are thoroughly linked. This will enable manufacturing firms to achieve a more efficient, flexible, and sustainable manner with consistently excellent quality and minimum cost [29]. Manufacturing enterprises must provide a greater return on investment while minimizing their negative effects on the environment. They must also create a welcoming workplace for those who value cooperation, education, and the growth of their competencies.

In this regard, this study aims to identify various digital technologies that can be effectively adopted in the manufacturing sector to achieve sustainable development. A total of eight digital technologies were discussed for their application in the manufacturing sector. For a better understanding of digital technologies in manufacturing, a schematic diagram is presented and is shown in Fig. 1.

2 Applications of Digital Technologies in the Manufacturing Sector

2.1 Embedded Systems in the Manufacturing Sector

Research on sustainable manufacturing processes, particularly for embedded systems, has many sides and requires both long-term analysis and immediate action. For instance, there are numerous potential advantages of embedded systems in the

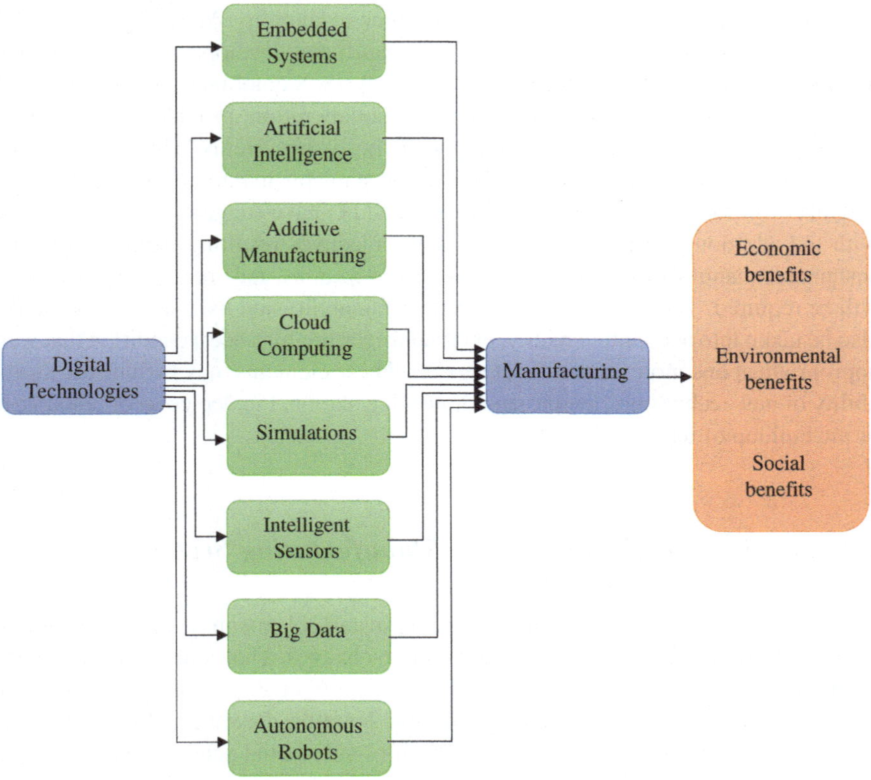

Fig. 1 Digital technologies in manufacturing

domain of sustainable production processes. However, logically centralized and multi-participative systems in the application of embedded systems are currently not methodologically meant to help with various critical socioeconomic and environmental issues. Trentesaux and Giret [28] for instance, suggest a green manufacturing initiative in its entirety. The goal of green manufacturing initiatives is to give a practical generic notion that is simple to adapt, enumerate, and apply. Additionally, the goal of the green manufacturing initiative is also to encourage academics to design sustainability-tailored production operations control systems, whether holistic or with a multipronged approach, with respect to embedded systems. Carvalho et al. [4] suggest sustainable manufacturing with a perspective of Industry 4.0, wherein collaboration as a tool for achieving sustainability in manufacturing is discussed with a perspective on embedded systems. The authors further recommended implementing scientific studies that highlighted the advantages of the new industry model, such as enhanced product life cycles. For example, the production that continues to operate in an integrated manner with the use of technologically advanced application areas is allied to the foundations of a sustainable development application in industry, such as decentralization, configuration management, interoperability, and many others. This leads to even more flexibility in natural resource availability. Mennenga et al. [22] suggests that System of Systems (SoS) is a consequent development of continuous and increased digitization of technical systems. An SoS combines the resources and capabilities of its subsystems and, via clever collaboration, provides greater functionality than the combined functionality of those subsystems. The authors have attempted to address the relationship between sustainable manufacturing, Life Cycle Engineering (LCE), and SoS. Garetti & Taisch [11] have suggested that to comply with the additional restrictions and the new objectives resulting from the goal of sustainable manufacturing, new product kinds, operations, and organizational models will be required. The dynamic character of sustainability and its components should also be taken into account; in reality, external forces might cause a condition that was appropriate at one moment to alter at another. These elements may include the availability of new technology, the passing of new legislation, the depletion of resources, or the buildup of undesired goods.

2.2 Artificial Intelligence in the Manufacturing Sector

The minimal fundamental needs for advancing sustainable manufacturing techniques in any industry are innovation and adaptation to changes. The same has been significantly influenced by artificial intelligence (AI). Cioffi et al. [5] have suggested that several AI-based solutions, including machine learning, have been founded in the sector to achieve sustainable goals as a result of substantial scientific study on the subject. Because of this, the authors proposed a review of the academic studies on the application of machine learning (ML) and AI in the industry. The increase in American publications and increased interest after the introduction of Industry 4.0 was a fascinating discovery. Bag et al. [2] have suggested the importance of big data analytics-powered AI and its growth over recent history. The authors contended that there was limited research concerning the requirements for which manufacturing firms have adopted big data-based AI for their sustainable manufacturing processes. The author contributed to the numerical validation of the theoretical framework. They further clarify the role of organizational practices with regard to the resource base and their effects on the adaptation of artificial intelligence powered by big data analytics. This affects the functionality of sustainable manufacturing and the circular economy while being restrained by organizational flexibility and industry dynamism. Bag and Pretorius [1] additionally, have further stressed that there has been very less research on the impact of Industry 4.0 technology implementation on sustainable manufacturing methods. The article concludes that smart products and new technologies which have embedded AI could have significant economic, social, and cultural contributions toward reaching sustainability in manufacturing practices.

2.3 Additive Manufacturing in the Manufacturing Sector

In the sphere of industrial processes, where significant amounts of energy and resources are utilized, cleaner production and sustainability are vital. Today, functional items with significant added value may be produced using additive manufacturing techniques like direct additive laser manufacturing. Bourhis et al. [3] have suggested that apart from studying the benefits of additive manufacturing processes, it is also important to study the environmental impact of those manufacturing processes with a motive of easier acceptability of sustainable manufacturing processes across the world. This study introduces a novel approach to environmental impact assessment that considers all consumed fluxes (material, fluids, and power). This methodology combined a precise assessment of flow consumption in the machine with the broad perspective needed for a sustainable strategy. This has been supported by Frăţilă and Rotaru [10], wherein they further add that the sustainability of additive manufacturing processes must be evaluated to facilitate their adoption and deployment in the sector, given the significance of sustainable practices in manufacturing processes. Furthermore, by considering the ecological factors when a product is being manufactured, the producers may increase their profitability and competitiveness. Machado et al. [20] further elaborate that the future of manufacturing will be greatly influenced by emerging technologies like additive manufacturing, which will allow businesses to make complex goods and components more effectively and sustainably. Companies, however, are finding it difficult to recognize and realize the technology's full sustainability potential. Creating a checklist to evaluate the sustainability initiatives of additive manufacturing uptake and exploitation was the goal of the study. Siva Rama Krishna and Srikanth [25], further, add that the basic concept of sustainable manufacturing emancipates manufacturing practices that have the least impact on the environment. It also refers to actions that may be taken again without endangering future generations' access to resources. The authors have assessed how sustainable manufacturing is impacted by subtractive and additive manufacturing techniques.

2.4 Cloud Computing in the Manufacturing Sector

Internet of Things (IoT) and cloud computing (CC) have received a lot of attention recently since they offer a new approach for intelligent observation and connection from "man-to-man," "man-to-machine," and "machine-to-machine relationships." Tao et al. [8] elaborated that the applications of the IoT and CC technologies in production are studied in order to accomplish the complete sharing, free circulation, on-demand usage, and optimal allocation of diverse industrial resources and capabilities. Liu et al. [18] elaborate on cloud robotics which is a combination of cloud computing and robotics. The paper professes the concept of robots as services and its impact on the development of robots in the future. Industrial cloud robotics (ICR) is advised to connect automated robotic resources worldwide to provide ICR services globally due to its benefits of easy access, resource sharing, and lower pricing. Industrial productivity greatly rises as a result of ICR. Future directions for the manufacturing sector are being investigated, including cloud manufacturing and sustainable production. The key to industrial sustainability is ICR's energy consumption optimization. A paradigm shift in manufacturing is being brought about by industrial value creation, according to research by Li et al. [17]. This paradigm shift is being aided by modern information and communications technology like the internet, computer systems, Internet-of-Things, cloud services, and big data. Additionally, they are also influenced by the growing international resource limitations and business diversity, the diversity of individual needs, and the long-term goals of environmental sustainability.

2.5 *Simulation in the Manufacturing Sector*

The industrial sector has benefited from the development of digital engineering tools and techniques. However, a variety of system parameters must be jointly optimized to build a sustainable production system. Heilala et al. [13] studied that, to enhance efficiency, and reconcile environmental limitations, a comprehensive simulation tool may be used during the system design stage. A sustainable production system must consider lean manufacturing, waste and loss of production detection and eradication, and environmental concerns. Shao et al. [24] profess that designing manufacturing systems with a lower environmental impact is essential for achieving sustainable production. The success of achieving sustainability depends on modeling and simulation (M&S), which can foresee the results of putting particular facilities, processes, and production activities in place. The information must be shared with a range of production systems, apps, and databases in order to implement M&S for sustainable manufacturing. Additionally, M&S offers the right concepts and techniques for creating algorithms that accurately convert data. Gbededo et al. [12] offer a conceptual framework that helps in the development of an extensive tool of analysis. It is based on a simulation that combines objectives that support the development of viable product development with techniques that are focused on the in-depth quantitative analysis of all pertinent sustainability dimensions. Lee et al. [16] have proposed a fresh method for simulating the sustainability initiatives of manufacturing. It is suggested that an information model serves as a neutral framework for data interchange. In order to achieve sustainable goals while maintaining productivity via simulation, the manufacturing industry must understand certain important obstacles.

2.6 *Intelligent Sensor in the Manufacturing Sector*

A monitoring system can be installed in a manufacturing process for a variety of reasons. Modern manufacturing equipment must be adaptable, durable, and function with the least amount of human interaction possible. Error-free machine operation is also required. Stavropoulos et al. (2013) suggest that in a machining setting, tool breakage is a primary source of unscheduled halt and is expensive in terms of both wasted time and capital. Additionally, a significant element in modern production is the products' environmental footprint. Manufacturing companies are under increased pressure to cut their energy use for cost-saving and environmental reasons because of the rising cost of energy and the current sustainability movement. Today, industrial sectors' growth and sustainability are driven by the monitoring and management of their production processes. Today, manufacturing industry progress and sustainable practices are being driven by the monitoring and management of production processes. The controllability and dependability of machining operations can be improved by adaptive control systems.

2.7 *Big Data in the Manufacturing Sector*

Future-state manufacturing sometimes referred to as "smart manufacturing," has the potential to significantly alter every element of operations by reducing energy and material usage, optimizing sustainability, and allowing a more digitalized, futuristic manufacturing environment. Kumar et al. [14] have expressed that condition-based maintenance (CBM) optimization improves regular maintenance and increases forecast accuracy in the big data analytics framework. CBM would increase equipment dependability and lower maintenance costs by using condition monitoring and forecast information effectively. Dubey et al. [6] have suggested a conceptual framework that captures big data functions utilizing components discovered through the reduction of acquired data. This is followed by the step of testing this paradigm using heterogeneous, diversified, voluminous, and high-velocity data. Finally, it concludes with an emphasis on the significance of academics and practice. According to Stock and Seliger [27], the challenge of modern globalization is to meet the continually increasing demand for capital and consumer goods while also ensuring the social, ecological, and economic sustainability of human life. Machado et al. [21] have made an effort to address how sustainable exploration is making an impact on the development of the industry 4.0 agenda and for a wider set regarding links across both Industry 4.0 and Sustainable Production. This has been done by mapping and summarizing current research attempts, followed by identifying research topics, and areas for research and development.

2.8 *Autonomous Robots in the Manufacturing Sector*

The incorporation of autonomous robots is a key factor in current manufacturing systems that make production operations more efficient and geared toward the automobile industry. Using functional qualities, structural data, activities, and process conditions, a unitary sustainable manufacturing or production capacity of the autonomous robot model was built [30]. To offer a differential representation for both the static and dynamic properties of sustainable manufacturing capacity, a hybrid logic description technique is illustrated. Energy usage during the autonomous robotic process is presented in portions using an interval-state description approach. Three categories of rules—stability, energy usage, and production capacity—are developed based on the built-in model for reasoning autonomous robots' capability. Franciosi et al. (2020) have emphasized that manufacturing companies have a significant influence on the most important pillars of sustainability (economic, environmental, and social), and they are now key players in the development of sustainable value. Enyoghasi & Badurdeen (2021) further adds that the need for achieving sustainability is driven by the scarcity of resources and the harmful effects that traditional manufacturing has on society and the environment. Integration of the product, process, and system is essential for sustainable manufacturing, taking into account the interconnected sustainability consequences. This was followed by a focused approach to the fundamental technologies of perception, cognition, decision-making, execution, and evolution by the authors [19]. This research also established a framework for intelligent production and disassembly, demonstrating the potential applications of various AI technologies.

3 Challenges to Adopt Digital Technologies in the Manufacturing Sector

From the literature review, it is found that the majority of researchers have agreed that there are still many challenges in adopting digital technologies in the manufacturing sector [23]. Digital technologies provide numerous benefits in manufacturing such as higher productivity, mass personalization, increased machine life, higher quality, and low cost of production. Despite several advantages, there are several challenges in adopting digital technologies in the manufacturing sector and some of them are as follows [23]:

- Unskilled workforce
- Insufficient financial resources
- Lack of data security
- Low degree of standardization
- Lack of vision
- Complexity in value chain integration
- Lack of infrastructure
- Lack of digital strategy
- Lack of clarity on economic benefits.

4 Conclusions and Future Research Directions

Digital technologies are widely being used in tremendous applications in the manufacturing sector, more specifically, in machine health prediction, demand forecasting, personalized product designs, real-time process monitoring, and many more. This study discusses the applications of digital technologies in manufacturing to gain sustainable benefits. Eight different digital technologies namely embedded systems, artificial intelligence, additive manufacturing, cloud computing, simulation, intelligent sensor, big data, and autonomous robots have been discussed for their adoption in manufacturing. This study would benefit industry practitioners and researchers to identify the scope of digital technologies in the manufacturing sector to make it digital manufacturing thereby making the process sustainable. Based on the literature studies, the following future studies are suggested:

- Analyzing the challenges in adopting digital technologies more specifically for SMEs.
- Performance assessment of digitization in manufacturing firms.
- Adoption of digital technologies for multiple sourcing in the manufacturing supply chain to minimize the disruptions in the supply chain.

References

1. Bag S, Pretorius JHC (2022) Relationships between industry 4.0, sustainable manufacturing and circular economy: proposal of a research framework. Int J Organiz Anal 30(4):864–898. https://doi.org/10.1108/IJOA-04-2020-2120
2. Bag S, Pretorius JHC, Gupta S, Dwivedi YK (2021) Role of institutional pressures and resources in the adoption of big data analytics powered artificial intelligence, sustainable manufacturing practices and circular economy capabilities. Technol Forecast Soc Change 163:120420. https://doi.org/10.1016/j.techfore.2020.120420
3. le Bourhis F, Kerbrat O, Hascoet J-Y, Mognol P (2013) Sustainable manufacturing: evaluation and modeling of environmental impacts in additive manufacturing. Int J Adv Manuf Technol 69(9–12):1927–1939. https://doi.org/10.1007/s00170-013-5151-2
4. Carvalho N, Chaim O, Cazarini E, Gerolamo M (2018) Manufacturing in the fourth industrial revolution: a positive prospect in sustainable manufacturing. Procedia Manuf 21:671–678. https://doi.org/10.1016/j.promfg.2018.02.170
5. Cioffi R, Travaglioni M, Piscitelli G, Petrillo A, de Felice F (2020) Artificial intelligence and machine learning applications in smart production: progress, trends, and directions. Sustainability 12(2):492. https://doi.org/10.3390/su12020492
6. Dubey R, Gunasekaran A, Childe SJ, Wamba SF, Papadopoulos T (2016) The impact of big data on world-class sustainable manufacturing. Int J Adv Manuf Technol 84(1–4):631–645. https://doi.org/10.1007/s00170-015-7674-1
7. Enyoghasi C, Badurdeen F (2021) Industry 4.0 for sustainable manufacturing: opportunities at the product, process, and system levels. Resour Conserv Recycl 166:105362. https://doi.org/10.1016/j.resconrec.2020.105362
8. Tao Fei, Cheng Ying, Da Li Xu, Zhang Lin, Li Bo Hu (2014) CCIoT-CMfg: cloud computing and internet of things-based cloud manufacturing service system. IEEE Trans Ind Inform 10(2):1435–1442. https://doi.org/10.1109/TII.2014.2306383

9. Franciosi C, Voisin A, Miranda S, Iung B (2020) Integration of I4.0 technologies with maintenance processes: what are the effects on sustainable manufacturing? IFAC-PapersOnLine 53(3):1–6. https://doi.org/10.1016/j.ifacol.2020.11.001
10. Frăţilă D, Rotaru H (2017) Additive manufacturing—A sustainable manufacturing route. MATEC Web Conf 94:03004. https://doi.org/10.1051/matecconf/20179403004
11. Garetti M, Taisch M (2012) Sustainable manufacturing: trends and research challenges. Prod Plann Control 23(2–3):83–104. https://doi.org/10.1080/09537287.2011.591619
12. Gbededo MA, Liyanage K, Garza-Reyes JA (2018) Towards a life cycle sustainability analysis: a systematic review of approaches to sustainable manufacturing. J Clean Prod 184:1002–1015. https://doi.org/10.1016/j.jclepro.2018.02.310
13. Heilala J, Vatanen S, Tonteri H, Montonen J, Lind S, Johansson B, Stahre J (2008) Simulation-based sustainable manufacturing system design. In: 2008 winter simulation conference, pp 1922–1930. https://doi.org/10.1109/WSC.2008.4736284
14. Kumar A, Shankar R, Thakur LS (2018) A big data driven sustainable manufacturing framework for condition-based maintenance prediction. J Comput Sci 27:428–439. https://doi.org/10.1016/j.jocs.2017.06.006
15. Kumar V, Sezersan I, Garza-Reyes JA, Gonzalez EDRS, AL-Shboul MA (2019) Circular economy in the manufacturing sector: benefits, opportunities and barriers. Manag Decis 57(4):1067–1086. https://doi.org/10.1108/MD-09-2018-1070
16. Lee JY, Kang HS, do Noh S (2014) MAS2: an integrated modeling and simulation-based life cycle evaluation approach for sustainable manufacturing. J Clean Prod 66:146–163. https://doi.org/10.1016/j.jclepro.2013.11.029
17. Li K, Zhou T, Liu B (2020) Internet-based intelligent and sustainable manufacturing: developments and challenges. Int J Adv Manuf Technol 108(5–6):1767–1791. https://doi.org/10.1007/s00170-020-05445-0
18. Liu J, Xu W, Zhang J, Zhou Z, Pham DT (2016) Industrial cloud robotics towards sustainable manufacturing. In: Materials; biomanufacturing; properties, applications and systems; sustainable manufacturing, vol 2. https://doi.org/10.1115/MSEC2016-8733
19. Liu Q, Liu Z, Xu W, Tang Q, Zhou Z, Pham DT (2019) Human-robot collaboration in disassembly for sustainable manufacturing. Int J Prod Res 57(12):4027–4044. https://doi.org/10.1080/00207543.2019.1578906
20. Machado CG, Despeisse M, Winroth M, da Silva EHDR (2019) Additive manufacturing from the sustainability perspective: proposal for a self-assessment tool. Procedia CIRP 81:482–487. https://doi.org/10.1016/j.procir.2019.03.123
21. Machado CG, Winroth MP, Ribeiro da Silva EHD (2020) Sustainable manufacturing in Industry 4.0: an emerging research agenda. Int J Prod Res 58(5):1462–1484. https://doi.org/10.1080/00207543.2019.1652777
22. Mennenga M, Cerdas F, Thiede S, Herrmann C (2019) Exploring the opportunities of system of systems engineering to complement sustainable manufacturing and life cycle engineering. Procedia CIRP 80:637–642. https://doi.org/10.1016/j.procir.2019.01.026
23. Raj A, Dwivedi G, Sharma A, Lopes de Sousa Jabbour AB, Rajak S (2020) Barriers to the adoption of industry 4.0 technologies in the manufacturing sector: an inter-country comparative perspective. Int J Prod Econ 224:107546. https://doi.org/10.1016/j.ijpe.2019.107546
24. Shao G, Bengtsson N, Johansson B (2010) Interoperability for simulation of sustainable manufacturing. In: Proceedings of the 2010 Spring simulation multiconference on—SpringSim', vol 10, p 1. https://doi.org/10.1145/1878537.1878595
25. Siva Rama Krishna L, Srikanth PJ (2021) Evaluation of environmental impact of additive and subtractive manufacturing processes for sustainable manufacturing. Mater Today: Proc 45:3054–3060. https://doi.org/10.1016/j.matpr.2020.12.060
26. Stavropoulos P, Chantzis D, Doukas C, Papacharalampopoulos A, Chryssolouris G (2013) Monitoring and control of manufacturing processes: a review. Procedia CIRP 8:421–425. https://doi.org/10.1016/j.procir.2013.06.127
27. Stock T, Seliger G (2016) Opportunities of sustainable manufacturing in industry 4.0. Procedia CIRP 40:536–541. https://doi.org/10.1016/j.procir.2016.01.129

28. Trentesaux D, Giret A (2015) Go-green manufacturing holons: a step towards sustainable manufacturing operations control. Manuf Lett 5:29–33. https://doi.org/10.1016/j.mfglet.2015.07.003
29. Wang S, Wan J, Li D, Zhang C (2016) Implementing smart factory of industrie 4.0: an outlook. Int J Distrib Sens Netw 12(1):3159805. https://doi.org/10.1155/2016/3159805
30. Zhao Y, Liu Q, Xu W, Wu X, Jiang X, Zhou Z, Pham DT (2017) Dynamic and unified modelling of sustainable manufacturing capability for industrial robots in cloud manufacturing. Int J Adv Manuf Technol 93(5–8):2753–2771. https://doi.org/10.1007/s00170-017-0634-1

Index

J. Kandasamy et al. (eds.), *Progress in Sustainable Manufacturing*, Management and Industrial Engineering, https://doi.org/10.1007/978-981-99-0201-9

The manufacturer's authorised representative in the EU is Springer Nature Customer Service Centre GmbH, Europaplatz 3, 69115 Heidelberg, Germany. If you have any concerns regarding our products, please contact ProductSafety@springernature.com

Printed and bound by CPI Group (UK) Ltd, Croydon, CR0 4YY
30/09/2026
02225314-0001